Determinants of Success of Collective Action
on Local Commons

T0316398

DEVELOPMENT ECONOMICS AND POLICY

Series edited by Franz Heidhues and Joachim von Braun

Vol. 49

PETER LANG

Frankfurt am Main · Berlin · Bern · Bruxelles · New York · Oxford · Wien

Determinants of Success of Collective Action on Local Commons

An Empirical Analysis of Community-Based Irrigation Management in Northern Ghana

Kadir Osman Gyasi

PETER LANG
Europäischer Verlag der Wissenschaften

Bibliographic Information published by Die Deutsche Bibliothek
Die Deutsche Bibliothek lists this publication in the Deutsche Nationalbibliografie; detailed bibliographic data is available in the internet at <http://dnb.ddb.de>.

Zugl.: Bonn, Univ., Diss., 2005

This research was made possible by
the Robert Bosch Foundation under the Devolution
and Community Management Project at ZEF, and the
International Water Management Institute (IWMI).

D 98
ISSN 0948-1338
ISBN 3-631-54084-1
US-ISBN 0-8204-7738-9

© Peter Lang GmbH
Europäischer Verlag der Wissenschaften
Frankfurt am Main 2005
All rights reserved.

Printed in Germany 1 2 3 4   6 7

www.peterlang.de

To my parents Alhaji Ansomana and N'ABiba, who never had the opportunity to enjoy formal education but realized its importance by getting me this far.

PREFACE

Devolution of natural resource management from government agencies to user groups has been gaining importance in the contemporary development agenda. This process is in response to inefficiencies of the state in managing common-pool resources, coupled with deteriorating fiscal positions of governments, and the growing move towards decentralization and participatory approaches promoted by donor agencies. This paradigm shift in natural resource management has been much pronounced in the developing world where farmer participation in the management of irrigation systems through water users' associations has been widely practiced. In spite of the great optimism for the potential of users' associations to solve the problems of commons management, the actual outcomes of the devolution programs in various countries have been mixed, and often the stated objectives of achieving positive impacts on resources productivity, equity and sustainability are not met. Against this background, Kadir Osman Gyasi conducted this empirical study in an attempt to explain why communities differ in terms of distributional and environmental (sustainability) outcomes of devolution programs.

Indeed, past analysis of institutional arrangements in farmer managed irrigation systems have led to proposed design principles for successful local management of natural resources. However, very few studies have vigorously tested these proposals in relation to irrigation management. While most of the existing studies have been carried out in Asia and Latin America, very little is known about the outcomes of local management of irrigation systems in Africa. This study by Kadir Osman Gyasi bridges this knowledge gap by empirically testing the existing hypotheses on community-managed irrigation schemes in northern Ghana.

The study presents a vivid account (both of the strengths and the weaknesses) of the endogenous traditional institutions in natural resource management in the study region and highlights the changing relationship between the state and users of the resource system in the operation and maintenance of public irrigation schemes. He also outlines the institutional and socioeconomic backgrounds that favor local management of irrigation systems.

Gyasi's empirical analysis of the irrigation communities in northern Ghana points to considerable collective action potential of rural communities and provides empirical support for the crucial roles of group size and social heterogeneity for the success of a user group's collective action, which have attracted a considerable debate in the literature on collective action and community management of natural resources.

Prof. Dr. Franz Heidhues
University of Hohenheim
Stuttgart

Prof. Dr. Joachim von Braun
I.F.P.R.I
Washington, D.C.

ACKNOWLEDGEMENT

I would like to express my sincere gratitude to Prof. Dr. Klaus Frohberg for supervising this work and his confidence in me. His encouragement and guidance are very much appreciated. I would like to also thank Prof. Dr. Holm-Müller for accepting to be the co-supervisor of this dissertation at a very short notice. Prof. Holm-Müller will forever be remembered for her contributions.

I am deeply indebted to Dr. Stephanie Engel, without whom this thesis would not have come about. I am very grateful for this, but more so her sustained interest, encouragement, guidance and critical but valuable comments have been very important for the successful completion of this research work. I wish to thank the Robert Bosch Foundation, Germany for providing financial support for the study. The International Water Management Institute (IWMI) is acknowledged for financially supporting the field work of this study. My sincere gratitude goes Dr. Abdul Kamara of IWMI for his invaluable comments and suggestions. I am grateful to Dr. A. B. Salifu (Director), Dr. V. Clottey, Luke Abatania, James Kombiok, Baba Braimah, Cecil Osei, Dr. Y. Alhasan, Dr. M. Fosu, and other colleagues at the Savanna Agricultural Research Institute (SARI) of the Council for Scientific and Industrial Research (CSIR), for their support and encouragement. I am equally grateful to Mr. Roy Ayariga, Mr. Edward Andanye and other staff members of the LACOSREP, Bolgatanga for their invaluable assistance to this study. I also thank the enumerators who carried out the survey and the farmers for talking to us. I am indebted to Thompson Abagna for his diligence in the conduct of the survey.

I am very grateful to the CSIR for granting me a study leave. I gratefully knowledge the Center for Development Research (ZEF), University of Bonn, where I undertook this study as a PhD student, for the enviable academic environment and for the career development opportunities I was exposed to. Dr. Gunther Manske (Coordinator of the International Doctoral Program, ZEF) is also acknowledged for his guidance and administrative support.

I am very grateful for the excellent support I had from the group of Ghanaian students at ZEF. Thomas Bagamsah, Nathaniel Howard, Banabas Amisigo, Lansah Abdulai, Dilys Sefakor Kponkor and Dr. Isaac Osei-Akoto, I am so grateful for your support and company. May the Almighty God bless you all. I am equally thankful to Dr. Khalid Yousif Khalifalla for his comments on my work. My thanks also go to my office mates and group members for the working relationship we enjoyed over the three years. Special thanks go to Ms Malanie Zimmermann for the excellent job done in proofreading the manuscript and translating this work into the German language. God bless you, Malanie.

Finally, I thank my parents, my brothers as well as my lovely sister, Amina, for their prayers and moral support throughout the years.

TABLE OF CONTENTS

LIST OF TABLES

LIST OF FIGURES

ABBREVIATIONS AND ACRONYMS

CV	Coefficient of Variation
DMC(s)	Damsite Management Committee(s)
DPMU	District Project Management Unit
FAO	Food and Agricultural Organization of the United Nations
GDP	Gross Domestic Product
GIDA	Ghana Irrigation Development Authority
GLSS	Ghana Living Standards Survey
GPRSP	Ghana Poverty Reduction Strategy Paper
GPS	Geographic Positioning Systems
GSS	Ghana Statistical Service
ICOUR	Irrigation Company of the Upper Region
IFAD	International Fund for Agricultural Development
IMT	Irrigation Management Transfer
IWMI	International Water Management Institute
LACOSREP	Land Conservation and Smallholder Rehabilitation Project
MoFA	Ministry of Food and Agriculture, Ghana
NGOs	Non-Governmental Organizations
NID	Normally and Independently Distributed
NORRIP	Northern Region Rural Integrated Development Program
NR	Northern Region of Ghana
OLS	Ordinary Least Squares
PCU	Project Coordination Unit
SSID	Small-Scale Irrigation Division
UER	Upper East Region of Ghana
URADEP	Upper Region Agricultural Development Program
UWADEP	Upper West Region Agricultural Development Project
UWR	Upper West Region of Ghana
VC(s)	Village Committee(s)
VIP	Village Infrastructure Project
WUA(s)	Water Users' Association(s)
ZEF	Zentrum für Entwicklungsforschung (Center for Development Research)

1 INTRODUCTION

1.1 Background and problem statement

Natural resources in general and local commons in particular are very important sources of livelihood for many people in developing countries. The sustainable management of these resources is therefore crucial for socio-economic well-being and environmental quality. State involvement in the management of natural resources has been justified because of market failure (especially externalities and natural monopolies[1]), and the strategic importance of the resources (Dasgupta, 1993). However, recognition of the inadequacy of public institutions in promoting sustainable resource management practices, particularly at the local level, coupled with ambiguous and sometimes negative sustainability and equity impacts of privatization provide arguments in favor of greater participation of communities in the management of local resources (Baland and Platteau, 1996; Meinzen-Dick et al., 1996; Rasmussen and Meinzen-Dick, 1995; Lawry 1990).

In many countries, institutional weaknesses and performance inefficiencies of public irrigation agencies have led to high costs of operation and maintenance of irrigation schemes. Poor maintenance culture and lack of effective control over irrigation practices have often resulted in the collapse of irrigation systems (Groenfeldt, 1997). Moreover, irrigation agencies have largely been unable to raise sufficient revenues from the collection of water charges to meet operational expenses. Deteriorating government fiscal positions in the face of mounting operation and maintenance costs of the irrigation agencies have provided the stimuli for many governments to adopt programs to devolve responsibility of irrigation management to user groups (Brewer, et al., 1999; Svendsen and Nott, 1997; Meinzen-Dick et al., 2002). This is also coupled with anecdotal evidence of success in traditional schemes in Africa, especially those that are constructed, operated and managed by local communities with limited external control or intervention (Kamara, et al., 2001). Indeed, these efforts towards the improvement of irrigation management performance are consistent with current tendencies, mainly driven by economic reform policies, to reduce the size and costs of governments by devolving responsibilities and activities to the local level (Shah et al., 2002; Meinzen-Dick and Knox, 1999; Kiss, 1990). Devolution or management transfer programs are also motivated by growing optimism that communities or user groups whose livelihoods depend on the resource may have greater incentives to effectively manage the resources to ensure efficiency, equity and sustainability (Shah, et al., 2002; Meinzen-Dick et al., 2002).

---

[1] The creation of irrigation facilities requires large and indivisible investment costs, creating a natural monopoly that a state can fill.

The concept of devolution in irrigation management implies a shift of responsibility and authority from state to nongovernmental institutions (including traditional institutions), and ranges from support for participation of farmers in irrigation management to irrigation management transfer (IMT). The aim is to increase the responsibility and authority of farmers at some or all levels and aspects of irrigation management, which involves the transfer of management from government (or its agency) to local level organizations controlled by water users, or in which the water users have a substantial voice (Svendsen et al., 1997; Svendsen and Nott, 1997; Meinzen-Dick et al., 1996). The form IMT takes, however, varies by context: cultural; climatic; topographic; macroeconomic, and depends on the scale and type of the irrigation system involved (Tural, 1995).

In general, devolution in irrigation management aims at correcting the excessive concentration of decision making, authority and power in the hands of state agencies and at promoting active involvement of local people and communities in the management of their resources (Rasmussen and Meinzen-Dick, 1995; Ostrom, 1990; Wade, 1988). In some instances, however, it remains debatable as to whether this is just a pretext to real ends, e.g. the reduction of constraints to public funding. The process often involves the formation of organizations of formal user groups known as water users' associations (WUAs)[2]. This derives from the need for formal rules and procedures when it comes to the assignment of use rights and the payment of fees (Knox and Meinzen-Dick et al., 2001). Considerable insights, which have helped in the design of WUAs, have been obtained from research into traditional small-scale farmer-managed irrigation schemes through the 1980s (Ostrom, 1990; Wade, 1988).

Evidence from successful devolution programs in some parts of the world has motivated a lot of donor support for small-scale irrigation schemes under community management in Ghana. Leading this initiative, the International Fund for Agricultural Development (IFAD), through its assisted regional agricultural development projects, has funded the construction and rehabilitation of communally managed dams in the Upper East Region under the Land Conservation and Smallholder Rehabilitation Project (LACOSREP), with ownership rights and responsibilities transferred to beneficiary communities. The success story of LACOSREP has engendered the replication of this model by other NGO and donor funded programs with irrigation components in northern Ghana.[3] At the same time, the existing large-scale irrigation schemes (under public irrigation agencies) have increasingly devolved management rights

---

[2] FAO (1982:8) defines WUAs as organizations of water users that manage, allocate and distribute water from a common source in the most efficient participative manner to benefit all the members.

[3] Examples include the IFAD funded Upper West Region Agricultural Development Project (UWADEP), and the Village Infrastructure Project (VIP) funded by the World Bank.

to water users of the schemes. Also, the Ghana Irrigation Development Authority's (GIDA) new policy direction emphasizes important roles for beneficiary communities, not only in the management but also in the planning, design and construction of the schemes. The idea is to instill a greater sense of ownership and to make it possible for the resource users to acquire some technical capacity in the operation and maintenance of the schemes.

Experiences with the growing promotion of community-based irrigation management suggest that these institutions may be successful not only in promoting effective management and sustainable levels of maintenance, but also in contributing to an equitable distribution of benefits derived from the schemes. Theoretical advantages of resource management by users are well documented in a number of studies (e.g., Baland and Platteau, 1996; Bardhan, 1993; Bromley, 1992; Ostrom, 1992a), and often suggest optimism for the potential of user associations to solve problems of natural resource management (Meinzen-Dick, 2002). However, WUAs do not operate in a vacuum and are themselves affected by several factors (technological, socio-economic, policy – including market, price and resource policies). Successful devolution, it is argued, requires effective local institutions that are supported by a clear public policy on local management of natural resources. In spite of this realization, devolution programs are sometimes promoted in environments where the prerequisites do not exist. Vermillion (1999) notes that because irrigation management reform is mostly inspired by financial pressures and often driven by donor imperatives, devolution or IMT policies tend to be adopted before strategies for implementation are identified.

The actual outcomes of devolution programs in various countries have been mixed, and the stated objectives of achieving positive impacts on resource productivity, equity, poverty alleviation, and environmental sustainability are often not met (Knox and Meinzen-Dick, 2001). While numerous examples of successful local management of irrigation exist in different parts of the world, there are also several cases of failure that sometimes lead to a complete system breakdown (Shah et al., 2002; Bardhan, 2000; Richards, 1997; Vermillion, 1997; Wade, 1988). Evidence of success of local management of irrigation schemes especially in the smallholder context remains limited (Shah et al., 2002). Furthermore, most of the new initiatives for forming WUAs and management strategies do not evolve from the traditional system. Instead, the structures of WUAs are largely imposed by government agencies and the donor community. The long-term sustainability of local institutions in communities where traditional hierarchies and informal institutions still play an important role often remains questionable. The mode of organization and the pattern of interaction within the organization can affect the resource management outcome (Rasmussen and Meinzen-Dick, 1995). Therefore, understanding the factors that affect community-based organizations, and the types of organizations that facilitate sustainable management of irrigation schemes, is essential for making improvements in the outcomes of irrigation management at the community level.

Two main strands of literature have examined the factors contributing to an effective community management of local resources: (i) game-theoretic models of community level cooperation in resource management, and (ii) socio-anthropological case studies. These approaches put forward many hypotheses on the determinants for successful community management of local resources.[4] Although some of these hypotheses have been tested in the management of resources such as rangelands (Kamara, 2001; McCarthy, Kamara and Kirk, 2001), very few rigorous analyses have tested them in irrigation management (e.g., Dayton-Johnson, 2000a, 2000b; Bardhan, 2000). Moreover, most of the existing studies have been carried out in Asia or Latin America, while very little information on the African situation exists. Indeed, past analyses of irrigation management in Africa have focused mostly on assessing the efficiency or profitability of different schemes without much emphasis on the factors that condition households' cooperation in collective management of irrigation schemes (see, e.g., Makombe et al., 2001; Kamara et al., 2001).

A deeper understanding of the outcomes of community management of irrigation schemes in Africa and the reasons why communities differ in terms of economic, environmental and distributional outcomes of irrigation management is required. This study generates an understanding of these issues by focusing on community irrigation management in the Upper East Region (UER) and the Upper West Region (UWR) in northern Ghana as a case study. These two regions are the poorest of the country's ten regions (IFAD, 1999) and are predominantly rural. Subsistence agriculture is the main occupation of the majority of people. A high population density combined with highly erratic rainfall patterns and severe land degradation has led to annual food shortages and a vicious circle of poverty. To reduce the high risks associated with rainfed agriculture the Ghanaian government has embarked on an irrigation development policy, which is mostly focused on small and medium-scale irrigation schemes to be managed by local communities or user groups (Edig et al., 2002). However, previous experiences with community managed irrigation schemes have not always been positive. In fact, many schemes have severely deteriorated or broken down in the past due to insufficient maintenance and catchment area[5] protection (Dittoh, 1998). It is important to avoid a repetition of such problems to assure that policy objectives of poverty alleviation, livelihood improvements, and sustainable management will be achieved.

---

[4] Agrawal 2001, and Ramussen and Meinzen-Dick, 1995, provide summaries of the factors identified by some authors as affecting local management of natural resources.

[5] In the terminology used by LACOSREP, the term *catchment area* denotes the area immediately surrounding the reservoir. In the past, many irrigation schemes have severely deteriorated and often broken down as a consequence of siltation resulting from farming in the catchment area. 'No farming' rules and the planting of trees and grasses in the catchment area are considered crucial to sustainable scheme management.

1.2 Objectives and research questions

In this study we have analyzed the factors that determine success of community-based management of irrigation systems in northern Ghana, by testing the relative significance of hypothesized factors based on the literature on collective action and community management as well as country specific factors that condition the success of collective action in the management of common-pool resources. The study sought to achieve this by answering the following research questions:

1 To what extent does community management of irrigation systems improve outcomes (equity, maintenance efficiency, etc.) given the specific conditions of small farmers in northern Ghana?

2 Under what conditions are farmers most likely to participate in collective management of the irrigation schemes?

3 Which factors explain the observed differences in success of community-based management of irrigation systems and which policies can improve outcomes?

4 What are the desirable conditions for long-term sustainability of successful community-based irrigation management strategies, both in terms of sustainability in maintenance and sustainability of local institutions (e.g., water users associations)?

Addressing the above issues in northern Ghana is expected to contribute to a more successful, efficient and equitable agricultural development of the area, which is necessary to improve the welfare of the local population and reduce poverty and resource degradation. As noted in Dayton-Johnson (2000a), the systematic study of group performance in governing resource-use in communities can shed light on the determinants of success of collective action. It is hoped that the results of this study will help to bridge the knowledge gap that exists between theoretical predictions and actual results of irrigation management transfer to enhance the understanding of critical issues in the management of irrigation systems.

This case study is focused on northern Ghana, more importantly the Upper East and Upper West regions, but the research findings will hopefully find a useful application in other parts of the country with a similar agro-ecological and socio-economic setting.

1.3 Study overview

The rest of this book is organized into six chapters. Chapters 2 and 3 are descriptive and are meant to acquaint the reader with northern Ghana as well as irrigation management institutions in the study area. In chapter 2 we present background information on the region's physical environment, the structure of the local economy and the socio-political setting of the study area. The chapter also includes a discussion on traditional institutions in the study area and the roles these play in natural resource management. Chapter 3 focuses on irrigation development and examines the dynamics of irrigation management institutions in the study area. The concept of water users' association and how it has been organized for the management of community irrigation schemes in the study area is examined. Highlights of common problems and conflicts associated with local management are also presented in chapter three.

Data sources and sampling procedures are presented in chapter 4. The chapter also presents a description of the survey user groups and households. Resource (labor and cash) mobilization performance as well as maintenance activities carried out by the WUAs are discussed. Chapter 4 also highlights the multiple uses of the irrigation water and the potential effects of the benefits of the irrigation system on communal response to the maintenance of the schemes. Chapter 5 lays out the conceptual framework of the study. It begins with a review of the existing literature on commons management, with emphasis on factors that have been hypothesized to condition success of collective action. Though performance indicators are difficult to measure, the chapter outlines outcomes of community irrigation management that are used in the assessment of community performance. A theoretical framework highlighting the inefficiency in private provision of common good is used to throw some light on the potential of collective action for efficient management of irrigation schemes.

Chapter 6 presents empirical analyses of the outcomes of collective action for the maintenances of the community irrigation schemes. Questions about which factors determine successful levels of these outcomes are examined. The first part of the chapter examines collective action in general scheme management, covering distributives rules, rule conformance, quality of maintenance as well as their determinants. In the second part of the chapter, an agricultural household labor model is used to analyze the factors that influence household effort contribution to the collective maintenance of the community irrigation schemes.

Chapter 7 concludes the study by providing a summary, addressing policy implications and pointing to future research.

2 GEOGRAPHIC, SOCIOECONOMIC AND INSTITUTIONAL SETTING OF THE STUDY AREA

2.1 Introduction

The management of irrigation water is affected by physical as well as social characteristics of the areas where the schemes are situated. The impact of the natural and social environment on the behavior and activities of the participants is crucial for the long-term sustainability of the irrigation schemes. This chapter examines the physical, environmental, socioeconomic and socio-political characteristics as well as indigenous institutions of the study area to provide an understanding of the setting in which the analysis is placed.

2.2 Location and physical characteristics of northern Ghana

Northern Ghana, which this study limits itself to, is comprised of three administrative regions: Northern, Upper East and Upper West regions. The study area is situated between latitude $8° - 11° N$ and longitude $3°W - 0° 60E$. It borders Cote d'Ivoire to the West, Togo to the East and Burkina Faso to the North. In the South, the Black Volta and the Oti River roughly form its boundaries with Brong-Ahafo and the Volta Regions of Ghana respectively (Figure 2.1). Together, the three regions constitute about 41% of the total land area and about 18% of the total population of Ghana. Specifically, the Ghana 2000 census report (GSS, 2002) reveals that Northern Region (70,384 sq. km) is home to 1,820,806 people, whilst the Upper West (18,476 sq. km) has a population of 576,583. Upper East is the smallest of all the regions in northern Ghana representing about 3% (8,842 sq. km) of the land area, and has a population of 920,089. The Upper East Region is estimated to have the highest population density of all regions within Ghana (104.1 persons per sq. km compared with national figure of 77 persons per sq. km). The population densities for the Upper West Region and the Northern Region are 31.2 and 25.9 respectively. It is, however, estimated that over 80 percent of the population of northern Ghana live in rural areas.

Most of the study area is situated in the Guinea savanna ecological zone with the north-eastern portion approximating the Sudan savanna ecology. The natural vegetation is savanna woodland generally consisting of short deciduous, widely spaced, fire resistant trees and shrubs and short grasses. The natural vegetations particularly in the north-eastern portion have been seriously influenced by human activities and reduced to open parks where only trees species of economic value such as *Parkia filicoides* (dawa-dawa) and *Butryopernum parkii* (shea nut tree) and some Acacia species can be found growing, especially around dwelling areas (Dogbe, 1998). Indeed, annual bushfires have reduced the vegetation to what can be described as resistant tree savannah (Korem, 1985).

Figure 2.1: Map of Ghana showing the regions in northern Ghana

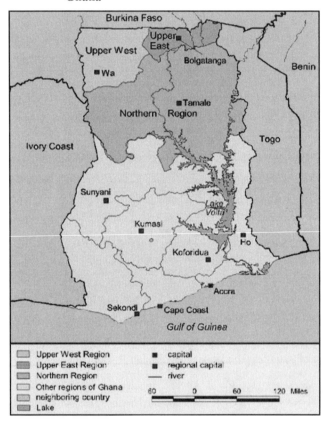

Soils in northern Ghana are inherently low in fertility[6], which is being worsened by soil erosion, continuous cropping and bushfires (Bruce et al., 1999, Abatania and Albert, 1993). However, the potential of agriculture in the area is high when normal rainfall is received. The majority of the active population is engaged in agriculture, but hash climatic conditions continue to make rain-fed agriculture, on which farmers depend, risky and a real challenge.

---

[6] Low organic matter contents and phosphorus reserves as well as occurrence of plinthite and erodible sandy top soils make the soils inherently infertile (see Asiamah, 2002).

Climatically, the area is characterized by wet and dry seasons. Rainfall is mono-modal often starting in April/May and ending in October with frequent dry spells during June and July. Its distribution is however erratic. Rainfall is highest in August or September (see Fig. 2 for a representative climatic data from Navrongo). An average annual rainfall of about 1000mm is received over the study area. Rainfalls of about 100mm, often exceeding the infiltration rate of the top soils, can be recorded within 24hr during the peak period. It is a common occurrence to experience both floods and droughts within the same rainy season. The region experiences average daily air temperatures ranging from about 35 Degrees Celsius in April to around 25 Degrees Celsius in August with corresponding low radiation and daily sunshine duration over the same period. The mean temperature over the area is about 29 °C.

The dry season (November to March) is characterized by dry harmattan (or north-east trade winds) from the Sahara. During the dry period, maximum daily air temperatures can reach 42° Celsius resulting in a very high evapotranspiration rate. The total evapotranspiration (in the vicinity of Navrongo) is about 2050mm, far exceeding the annual precipitation.

Figure 2.2: Mean annual rainfall and temperature

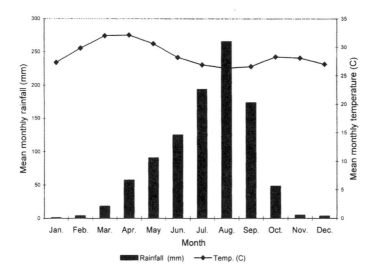

Source: ZEF database, 2003.

The area is drained by the Volta River and its tributaries. These large rivers as well as a number of small streams that flow through northern Ghana make most parts of the study area very suitable for irrigation development.

2.3   Economy

Agriculture is the main stay of the economy of Ghana in general (contributing 37% of GDP, three quarters of export earnings and a main source of employment for the majority of Ghanaians)[7] and northern Ghana in particular. This trend in the Ghanaian economy has persisted even before Ghana attained her independence from Great Britain.

The structure of the economy of the study area, in terms of economic participation of the population in the main sectors of the Ghanaian economy (agriculture, industry and services) is presented in Figure 2.3 below. Clearly, agriculture is the dominant sector of the economy of Northern Ghana, engaging 78 percent of the active population (75% in NR, 69% in UER and, 76% in UWR), while industry and services related sectors employ 10 percent and 16 percent respectively (GSS, 2002). This picture broadly underscores the potential Ghana has in using agriculture for economic development (income generation, poverty reduction, improved food security, etc.) of northern Ghana.

Figure 2.3: Activities of economically active population in northern Ghana.

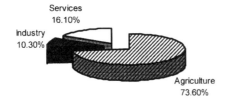

---

[7] In the year 2000, the contribution of Agriculture, Industry and Services to Ghana's GDP were 36.5, 25.2, 29.2 percent respectively, with Net Direct Taxes accounting for the remaining 9.1%. (The Budget Statement and economic policy statement of government of Ghana for the year 2000 financial year presented to parliament, Feb. 9, 2000).

Subsistence agriculture in particular is the main occupation of the majority of the population; producing agricultural products mainly for home use. However, household cash income is gained mainly from the sale of crops and livestock, and non-farm (economic) activities. Agricultural incomes represent over 71% of all rural household incomes in northern Ghana. Non-farm activities account for about 16% of household incomes whilst wage (mostly paid casual employment) and remittances constitute only a modest fraction of incomes as compared to farming (Table 2.1). Beside its dominant role in the composition of household income, agriculture is also vital for food supply in Northern Ghana (Asenso-Okyere et al., 2000).

Table 2.1: Percentage distribution of household annual income by source and regions in northern Ghana in the year 2002

Region	Components of Income						
	Wage	Agric	Non-farm	Rental	Remittances	Other	Total
Northern	11.4	58.6	23.2	3.0	3.0	0.8	100.0
Upper West	16.5	46.4	26.4	4.0	1.8	4.8	100.0
Upper East	12.9	66.6	15.0	2.7	4.4	1.3	100.0
Rural Savanna	6.4	71.4	16.0	2.0	3.0	1.2	100.0
National	21.9	35.5	30.0	1.9	8.5	2.1	100.0

Source: Ghana Statistical Service (GLSS 4), 2000

Indeed, agriculture in northern Ghana has the greatest potential to meet the country's food requirements, provide raw agro-industrial materials and export of agricultural commodities (World Bank, 1978). However, the region continues to be characterized by low productivity, as a result of the continuous use of rudimentary farm tools (hand-hoes), low soil fertility, and unfavorable and erratic climatic conditions. Mainly, agriculture in the study area (like in other parts of the country) is rainfed, relying on the seasonal and unpredictable rainfall that lasts from May to October. The uni-modal rainfall pattern prevailing in the region limits farmers to only one crop harvest per year. Crop failures and food shortages resulting from drought and flood, both occurring in the same year, are rampant especially in the Upper East Region of Ghana. Crop yields are low, generally limiting production to household food requirements, and increasing the vulnerability and food insecurity of farm households in the study area. The long dry period of six months that characterizes the study area is associated with idleness and lack of economic opportunities.

In spite of its climatic and capital constraints, northern Ghana makes an important contribution to the nation's food supply. The region accounts for over 40% of cereals and 63% of grain legumes produced in Ghana (MoFA, 1998). However, due to low factor productivity and for that matter very low agricultural production that characterizes the study area, the farming households universally live in poverty. Indeed, the study region remains the poorest and most deprived of all areas in Ghana. Poverty is widespread and food insecurity persistent. Evidence of growing and deepening poverty and intensification of vulnerability is widespread. Calorie intake is low and seasonal malnutrition prevalent especially among children (Ghana, 2002). Indeed, about nine out of ten people in UER (i.e., 88.2%), eight out of ten in UWR (83.9%) and seven out of ten in NR (69.2%) were classified as poor in 1999 (GSS, 2000).

In an attempt to boost food security and raise rural incomes, several small dams have been constructed in the study area to provide water for livestock and dry season gardening. The majority of these schemes, in addition to Ghana's largest irrigation schemes (Tono and Vea), are located in the UER where population density is high, land is scarce and, incidence and depth of poverty is high. Traditional dry season gardening, where farmers irrigate by fetching water from wells and streams using buckets and gourds, is also common in the study area. High value crops such as rice, onions, tomatoes, pepper, cabbages, lettuce, as well as traditional leafy vegetables are produced in the area during the dry season.

2.3.1 Farming systems

In the rainfed farming system two main types of farms exist in the study area: compound farms immediately surrounding or in the vicinity of the homestead, and those lying at a distance away from dwelling areas often referred to as "bush" or "far away" farms (Benneh, 1973). These systems of farming are also found in many tropical regions of the world where population pressure on land is high (see Ruthenberg, 1980). Continuous fragmentation of land through inheritance patterns, declining soil fertility and erosion have contributed greatly to land shortages, especially in the UER, leading to virtual disappearance of bush farms in the agricultural landscape in most parts of the region (Konings, 1996). Land fallowing extensively practiced in the traditional farming systems (especially on bush farms) as a means of restoring soil fertility has either been significantly reduced in length (of fallow period) or virtually eliminated in many places in the study area (Kranz et al., 1998). Compound farms are smaller in sizes and are cultivated annually to staples and vegetables. Often cereals such as millet, sorghum and maize intercropped with cowpeas are the main crops on the compound farms. Land use intensity is high on the compound farms and no fallowing is practiced (Clottey and Kombiok, 2000). These farms are usually fertilized with household refuse and livestock droppings.[8]

Bush farms continue to be popular in the Northern and Upper West Regions where population density is much lower and pressure on land is less severe. They are usually up to 20km away from settlements and larger than compound farms. A variety of crops are cultivated on bush farms with cash

---

[8] Farmers tether livestock around their houses and keep changing the positions (dynamic kraal) so that their droppings can spread evenly around the compound (Clottey and Kombiok, 2000).

crops having high incidence. Cotton, groundnut, soybeans, bambara beans, rice, late millet, sorghum and maize are generally cultivated on the bush farms. The bush farms are largely unmanured, and are usually cropped in rotation with bush fallows. Population pressure and increasing land scarcity is, however, limiting the fallow length, and effectively reducing the ability of the system to restore soil fertility naturally.

Cropping systems vary very little in the three northern regions. All the major cereals (millet, sorghum rice and maize) and legumes (groundnuts, cowpea, soybean, and pigeon) are cultivated in all the three regions. However, root and tuber crops (yam and cassava particular) feature prominently in the cropping systems of the Northern Region (Runge-Metzger, 1993), and to some extent Upper East, Region. Fafra (and sweet) potatoes are grown mainly in the Upper East Region. Sole cropping of rice and soybean is common on the far away farms.

Generally, field preparation and seeding of the rainy season crops start in April and May, while harvesting starts in October. Traditionally, the head of the compound makes cropping decisions, often in consultation with older sons. Early maturing varieties of cereals (e.g., early millet, in the Upper East Region) and legumes are cropped (usually on the compound farms and some lowlands) at the onset of the first rains to help bridge the hunger gap (a period between May and July when most household granaries are empty).

The method of farming in the study area has not undergone any significant modernization. Fertilizer application is generally low. Indeed, the most important input in the agricultural production process is labor. Mechanization is generally low and farm operations are mainly manually done using hoes. There is low usage of bullock traction and which is chiefly practiced on the bush farms (Hesse, 1998). There is growing use of tractor for land preparation only, especially in the areas where Cotton Companies operate.

Irrigated agriculture is playing an important role in poverty alleviation in the study area, especially in the Upper East Region. With almost seven months of dry season (October/November – April/May) and very little dry season employment, dry season gardening has become a very important source of income for many households. The major irrigated crops, tomato, pepper, and onion, are considered high value crops due to the associated high demand and prices. These crops are cultivated in almost all the schemes. In schemes closer to urban centers, exotic vegetables such as lettuce, cabbage, green pepper, and carrots are also of high value (Ayariga, 2002). Other crops that are intensively cultivated under irrigation but considered as low value (in terms of price) are okra and traditional leafy vegetables such as kenaf leaves (*Hibiscus cannabinus*), "alefu" (African spinach - *Amaranthus cuentus*), "bitto" (Roselle - *Hibiscus sabdariffa*) and "ayoyo" (African sorrel -*Corchorus olitorius*). Field preparation for the dry season gardening often begins in October. Most of the irrigated crops have a three months growing duration and often mature in February and March. However, the planting and harvesting times for the irrigated crops often depend on how timely the field (rainy season) crops are harvested.

Livestock rearing for traditional ceremonies and prestige has been a way of life of the people of northern Ghana. Cattle, small ruminants (sheep, goats and pigs), and donkeys are the major species of livestock that are kept. Households also keep poultry including guinea fowls. In particular, the rearing and keeping of cattle is regarded as a store of wealth and measure of social status. The number of cattle a family has is a visible sign of its prosperity and status (Dickson and Benneh, 1988). Cattle are kept as insurance against unexpected need, calamity or hardships, and as draft animals. Therefore, every family strives to possess and keep cattle. The small ruminants (sheep, goats, pigs) as well as poultry (chickens, guinea fowl, and pigeons) are kept for reasons of immediate economic security (Agyepong et al, 1999). These are sold in the local markets for cash to manage immediate cash needs. Having no sheep or goat is a sign of desperate poverty.

Donkeys are kept, particularly in the Upper East Region, for carting commodities to and from market centers. In fact, the use of animal traction is more prominent in the Upper East and West Regions, but to a limited extent. Apart from economic reasons, livestock also plays important roles in the sociocultural and religious lives of the people; for social activities such as marriages, funeral rites and religious sacrifices.

2.3.2 Non-farm activities

Under the conditions of limiting ecological factors including land degradation, low agricultural productivity and lack of monetary income in northern Ghana, farm households are forced to adopt non-farm ways of earning incomes to supplement the agricultural economy. The most popular strategy (as it is has been in the past) is inter-regional migration to urban centers, cocoa growing areas and mining industries in the south (Mensah-Bonsu, 2003; Adeifio-Shandoff, 1982). During the long dry season (November to April), the majority of farmers (except those engaged in dry season gardening) is put out of active economic activity. It is common to find people engaged in all kinds of crafts and secondary occupation to supplement their incomes to improve the quality of their lives. Men earn cash from petty trading or craft work and engage in seasonal and long-term labor migration (Tripp, 1987). In areas with limited opportunities for earning income, dry season activities focus around construction and maintenance of homesteads.

Hunting and fishing are common pastimes for men, especially during the dry season. Labor migration and economic diversification increasingly involve women in several non-farm activities to supplement household income. Some of the non-agricultural activities women do to support themselves and their families include weaving and basketry, petty trading, pottery, shea-butter extraction, groundnut oil processing, pito (local beer) brewing, as well as firewood collection. Small-scale gold mining and stone quarrying are other emerging avenues for people in some parts of the study area, though with destructive consequences for the already fragile environment.

2.3.3 Marketing

The majority of agricultural produce and inputs are sold through local periodic markets, which are usually held either once per week (e.g., six-day cycle in the Northern and Upper West Regions) or twice per week (three-day cycle in the Upper East Region). The largest product market is located in Tamale, the northern regional capital. Other important market places include Wa, Bolgatanga and Bawku. These markets also function as relay markets, attracting traders from Southern Ghana as well as from other parts of regions in northern Ghana. There are also several rural markets which serve as assembling points for surrounding villages, with a further level of village markets serving the immediate communities (Gyasi, 2001). Infrastructure for food markets is thin. Not all villages have markets. Where there are a number of markets in the same area, market days are staggered in such a way that no two important markets are held the same day.

Agricultural prices exhibit seasonal price fluctuations reflecting the prevailing climatic conditions. Food prices in the study area normally peak by mid July when food is scarce and are very low during harvesting. The irrigated crops are more affected by the seasonal variability in price due to the short shelve lives of the crops.[9] Prices of tomato for instance range from ¢200,000.00 per crate (mean weight of 60kg) in January to ¢30,000.00 at the glut period in March.

In the farm households, women are usually responsible for the marketing of crops whilst men undertake livestock trading. Products sold are usually similar throughout the rural markets. In the dam communities, important wholesale trades in the dry season are onions, tomatoes, and pepper. Farmers mostly carry their head-loads of the commodities to the market places. However, donkey/ox carts are becoming important means of transporting commodities to markets, especially in the urban and peri-urban areas in the Upper East and Upper West regions. Farmers travel between 6-8 km to attend the periodic markets. In addition to the formal community markets, informal markets are sometimes organized at the damsites during the dry season where traders from urban centers go to buy fresh vegetables during harvesting.

Market days also serve as social occasions where farmers from nearby communities appear in their best clothes to meet and make friends, and to exchange news and information, including technology uptake (Gyasi et al., 2002). Markets also play an important role in the marriage culture of the people.

---

[9] Crops are sold fresh, no processing takes place.

They serve as the grounds for courtship and sometimes elopement associated with many marriages among young Dagaabas in the Upper West Region. Indeed, the extent of the social significance of the market day is greater with larger village markets (Anyane, 1962).

A good road network is critical not only to facilitate movements of people but also agricultural marketing. Access to markets is difficult for most communities due to poor roads and lack of transport. Apart from the main Tamale-Bolgatanga-Paga trunk road, the area is generally served by poor road network. Roads to the rural communities are largely impassable during the rainy season. Many of the feeder roads linking the communities in the study area are often washed off or become too muddy to be used by the few vehicles that shuttle between the settlements, thus cutting off the rural settlements, where the bulk of food and agricultural raw materials are produced, from the urban and market centers.

2.4   Household organization and labor arrangements

Households in northern Ghana are predominantly organized in compounds in which individual members have their own round huts. They are usually made up of a man, his wives, married sons and their wives and his grandchildren. The oldest male member heads the compound. When the compound constitutes one production unit, the land belonging to the household is not divided. In the traditional compounds, land use decisions are taken by the head in consultation with male adults or the heads of smaller households forming an integral part of the larger compound (Mensah-Bonsu, 2003). Crops required to meet the household requirement are jointly produced and the harvest managed by the head.   Because the household head is traditionally responsible for the provision of the basic food needs of the compound, he can call upon members to provide labor in his farm. Apart from tending to the 'compound farm', members can have individual or private plots on which they cultivate mostly cash crops, the produce of which is reserved for exclusive use. The private plots provide cash income and additional food for the farmers and their immediate dependants to supplement what the compound head provides.

The household head is responsible for allocating labor within the household. He can command the labor power of all the young males on the family farm. All adult males are traditionally charged with the responsibility of providing the household with grains and so they concentrate on farming whilst adult females are responsible for the household maintenance task.   Where they do not have their own farms, adult females also participate actively in the household farms.

The household head can also mobilize labor from relatives, friends within the community, sons-in-law, and relatives living in other villages for specific activities (such as weeding and harvesting of crops), and serve food and beverages. The host is not obliged to repay but it is moral for him/her to reciprocate if anyone (especially of the participants) makes a similar call. However, the rising costs of providing food and drinks for the participants of this form of communal labor is dwindling the ability of many households (especially the poor ones) to mobilize communal labor.

It is noteworthy that the labor market in the study area is very thin. It is, however, a common practice for young people in the rural households to engage in non-market mechanisms of labor exchange. Mostly, the young people depend on group or exchange labor to reduce labor risk on their private plots. Group or exchange labor arrangements often involve a group of young people working in turn on the fields of each other (reciprocal exchange). It may start with the most senior or the one who first initiated the formation of the group. Variants of this arrangement exist, one of which is the "market day exchange" where the group activities take place on market days (typically twice a week). The member on whose plot the group is working is expected to provide food and drink for the day (Konings, 1996). In the group labor exchange arrangements, it is obligatory for whoever has benefited to repay when it comes to the turn of others. Persons who violate the moral contract incur a significant cost; apart from being thrown out of the group, the bad reputation gained would deprive him/her of future opportunities to enter into contracts with other village members.

Communal exchange labor is thus a form of cooperation in which farmers come together to perform specific activities for one another in turns. This kind of communal arrangement and cooperation lead to the development of norms, a network of associations or trust because of which people tend to cooperate for common welfare. Also important is the fact that the patriarchal household arrangement makes social mobilization for collective action easy as households play important roles in mobilizing members for collective activities.

2.5 Socio-political background

The social organization of the area can be better understood in the context of the historical and socio-political evolution of northern Ghana.

Northern Ghana consists of different socio-political entities and ethnic communities even today, as it was prior to the establishment of the colonial administration in the area. Eyre-Smith (1933, cited in Benning 1996), recounted that the whole area consisted of 3 main kingdoms (Mamprusi, Dagomba and, Gonja) who extended their rule over other clans, together with minor kingdoms (including Wala and Nunumba) and autonomous village/clan states, outside the jurisdiction of the political authorities of the kingdoms. The colonial administration in the then Northern Territory of the Gold Coast strengthened the kingdoms and also brought together the smaller and autonomous communities to build larger states to allow the establishment of efficient and effective forms of local government. Secular chiefs were imposed on the areas where chieftaincy was new and 'Tindanaship'[10] rule had prevailed, as it was necessary to find some sort of centralized political authority to deal with the colonial administration as part of the indirect rule the latter established (Benning, 1996). There was little resistance to these moves, as the introduced social order did not interfere much with the religion and land administration of the clan societies (ibid).

Ironically, apart from a few ethnic groups such as the Gonja, Sisala, Vagla Kasena, among others, most of the ethnic groups in Northern Ghana trace their roots to a common ancestry, the legendry Toha - the "Red Hunter" (Bacho, 2001). This group is commonly referred to as the Mole- Dagbani and comprises the Dagomba, Mamprusi, Dagaaba, Grune, Kusasi, Nunumba, Moshi, etc. Generally, there is no sharp linguistic or cultural boundary between most of the tribes nor any precise political or structural boundaries between neighboring tribes (Manukian, 1952). Thus, the socioeconomic organization and the culture of the groups have much in common.

To date the Chieftaincy institution has matured throughout the region, and each village is headed by a chief normally nominated from among the royal family. There are paramount chiefs who oversee traditional administrative communities mostly made up of villages that have a similar ethnic background. The village chiefs report all serious problems to the respective paramount chiefs to whom they owe allegiance. At the regional level there is a regional house of chiefs made up of paramount chiefs in each respective region.

Each village chief appoints elders (often including heads of various sections of the village) to help him in the administration of his community. The Chief and the Tindana (priest of the earth god) play important roles in the socioeconomic lives of the people. As the political head, the chief organizes the people for community development whilst his court assists the Tindana to enforce social norms and taboos governing the use and protection of the community's natural resources (Section 2.6). Indeed, beyond chieftaincy the concept of the clan epitomized by the Tindanaship is crucial in land administration in many parts of northern Ghana. The customary law of most parts of northern Ghana vests the allodial title to land in the Tindana (the descendant of the first settler) who holds it in trust for the people and performs sacrifices to the ancestors and land gods to seek protection and bountiful harvests for the people (Kasanga, 1995).

However, the advent of the colonial rule seemed to have transferred most of the political roles of the Tindana to chiefs, leaving him only as a spiritual head. It is common to find areas today where chiefs are usurping the roles of the Tindana in land administration (Kasanga, 1996). The present (post-colonial) political administrations have not helped the situation. Political leaders and government officials seem to have contributed to the marginalization and the fading roles of the Tindana in land administration as they often deal with the

---

[10] Tindana is the priest of the earth god. They belong to the family of the original settlers in the village and consequently the custodians of the village's earth (natural resources)

chiefs. Tindanas are often not consulted on issues which even include land administration.

In Ghana's present political dispensation, the decentralization policy provides political representation at the grassroots for local development. Administratively, northern Ghana has 24 District Assemblies (13 in Northern Region, 6 in Upper East Region and 5 in Upper West Region).[11] The decentralization program offers the local government authority, greater autonomy not only for mobilizing resources for development but also legislative powers to enact and enforce bye-laws in respect of the environment.

The major ethnic groups in the study area are predominantly patrilineal in inheritance. Christianity and Islam are the major religions practiced alongside traditional religions. Most social events, especially funerals and festivals, are often performed in the dry season when food is expected to be available and people are expected to have much time to participate in the activities. Such social functions could have implications for household labor availability for dry season gardening, a major source of livelihood, especially in the Upper East Region.

2.6 Indigenous institutions for natural resource management

Local institutions play very important roles in the sustainable management of natural resources. However, lack of recognition for the role of indigenous institutions in natural resource management had led to failures of many natural resource management initiatives in developing countries.

Often, communities that were affected by the resource system were not empowered to take part in the management of these resources, and no use is made of community knowledge, information on local factors, as well as the enforcement advantage, which external agents may not have (Salas, 1994; Warren, 1991; Chambers, 1988; Uphoff, 1986b). Besides, evidence from a number of empirical studies suggests that success or failures of collective action in communities largely depend on the presence or absence of social norms conducive for cooperation (Baland and Plateau, 1996; Platteau, 1994; Ostrom, 1990). This section examines the indigenous institutions in the study area, the roles they play in natural resource conservation, how they enhance cooperation and, how they can serve as entry points in the search for local initiatives for natural resource management.

---

[11] Additional districts have been added after our field survey (5 more were added to NR, 2 to UER , 2 to UWR)

2.6.1 Traditional institutions

North (1981) identifies institutions with constraints on behavior in the form of rules and regulations and enforcement procedures. In this way cultures become very important to the extent that moral and ethical behavioral norms aim at reducing enforcement cost (Levi, 1988).

Indigenous institutions symbolize established local systems of authority and other phenomena that emanate from the socio-cultural and historical processes of a given society. Local socio-political structures include procedures, local rules and norms, ethics and social sanctions designed to guide behavior of members of the community. They originate from local cultures, have firm roots in the past and reflect the knowledge and experience of the local people (Appiah-Opoku and Matowanyika, 1997). These indigenous institutions are often organized around traditional roles and systems of authority that guide ethics and practices of the community. Often, communities are small enough to allow perfect information flow among members who interact continuously. This promotes trust and a feeling of oneness and enhances cooperation among members whilst the mechanisms of social ostracism discourage misconduct. Tradition can thus play a significant role in the success of community-based resource management through social norms of behavior and established patterns of authority and leadership (Baland and Platteau, 1996).

Indigenous institutions in the study area that contribute to natural resource management can be described as social, religious, political and judicial institutions. Social and religious institutions broach the underlying principles of social ethics, norms and mores which become legitimized in such traditional structures as chieftaincy and indigenous court systems (see also Appia-Opoku and Hyma, 1999), which underlie the indigenous political and judicial systems (section 2.6.2)

Social institutions develop from kinship, clan and family relations as wells as social systems, including land tenure. Social institutions in the local societies are characterized by a network of relationships that even extend to the dead, unborn, kinship groups etc. This category emphasizes collective decision-making, communal ownership and control of resources, shapes individuals' preferences and nurtures trust and reliability which extends to mutual dealings.

On the other hand, indigenous religious institutions (overseen by the Tindana and a community of traditional priests), are typified by ancestral worship, divinity of nature and deities or gods. Religious institutions permeate all aspects of life, form the basis of morality and dictate the code of conduct in the traditional society (Appiah-Opoku and Hyma, 1999). Religious institutions nurture ethics in the traditional society, and in the case of natural resource management, for instance, inform the collective environmental wisdom expressed through observance of sacred beliefs and cultural practices. Successful experiences and positive attitudes towards resource management are conveyed through myths, customs, norms and sayings (Spradly and McCurdy,

1980). Indeed, local taboos and inhibitions are related to practices that are believed to defile the earth. For example, an indigenous belief that the spirit that dwells in the Earth (land, water bodies, trees, rocks, etc) has its own power which is "helpful" if appeased and harmful if offended, is a moral sanction against abuse of natural resources (Appiah-Opoku and Hyma, 1999). Cults of ancestors provide a stabilizing principle in which many conventions are grounded (Caldwell and Caldwell, 1987). The moral sanction in the foregoing example is re-enforced by the general belief among the traditionalists that the spirits of ancestors continue to observe the behavior of the living and reward those who act in accordance with social norms and punish those who exhibit deviant behaviors. These norms guide the respect for and maintenance of environmental quality and conservation of biodiversity.

2.6.2 Traditional authority and resource management

Traditional authority is derived from political institutions in the indigenous system that center on chieftaincy systems (including the council of elders), and Tindanaship (section 2.5). Political power is exercised through hierarchical levels of authority. Paramount chiefs appoint divisional and sub-divisional chiefs who in turn appoint minor chiefs and headmen, who exercise political control (including transactive decision-making on the use of resources in their division) and report to their respective paramount chiefs. The minor chiefs and headmen in turn report to the appropriate divisional or sub-divisional chiefs and so up the pyramid. (The divisional, sub-divisional and minors chiefs and headmen are custodians of all lands under their division particularly in the Northern Region where the traditional councils are centralized, as well as in parts of the Upper West and East regions where some chiefs have usurped the roles of the Tindana in land administration.) In many communities in northern Ghana, councils of elders and tindanas play important roles in political decision-making, land administration and enforcement of social norms and taboos with respect to the environment.

Native court systems symbolize the indigenous judicial institution at the family, community, village, and paramountcy levels (Appiah-Opoku and Hyma, 1999). These courts presided by the chief and elders mediate as well as settle internal disputes and litigations including those involving land and communal resource use. These native courts are at hand to assist the Tindana and indigenous religious leaders to enforce local taboos and prohibitions that are enacted to ensure ecological integrity and protect the sanctity of deities. Thus, the native courts provide legitimacy to local institutions for resource management.

Indeed, indigenous institutions are organized on the basis of traditional roles and systems of authority, experience and knowledge rooted in local culture and social values that are veritable in structures such as clans, chieftaincy and indigenous court systems. They embody practices and the traditional lifestyle of communities relevant for sustainable resource use, and prescribe social norms and mores for the protection and maintenance of communal resources. Sanctions for breaking these norms are varying but are punitive enough to deter rule breaking or failure to participate in communal action for resource management.

Although the adherence to traditional customs, religions, norms and taboos dominates the lifestyle of the people in the rural communities of the study area, the changing perceptions on the spiritual nature of land (especially in the land scarce areas), the declining authority and functions of the Tindana and changes that are occurring through globalization (market access, migration, population growth, etc) are adversely affecting the ability of the traditional systems to adjust and evolve practices that suit the emerging patterns of lifestyle and to provide self-sustaining institutions for managing natural resources.

2.6.3 Land tenure

As shown in the preceding sections, rural communities largely operate on the principles of customary law or belief systems in which tradition provides guidelines to legal and administrative processes outside the central government. These processes among others determine property rights, including land tenure. Understanding the land tenure system is essential for creating institutional mechanisms for managing communal resources. Land in the study area is perceived to be a spiritual entity, which cannot be owned by an individual. It is said to belong to all who have rights to it and the generations yet unborn. Land is therefore not sold, leased, pledged or rented. It is held in trust of custodianship for the community by the tindana.[12] Communities use features such as rivers, mountains, rocks, and special trees as frontiers to demarcate their lands. These are socially so recognized in each community that land disputes are rare. The tindana, usually the patrilineal descendant of the first family that settled at the place has spiritual authority over the land. Traditionally, the tindana is the spiritual caretaker of the land resources and holds it in trust for the community (Bakang and Garforth, 1998). Superimposed on the tindanaship is a traditional political authority, the chieftaincy system (section 2.5) which also holds significant influence in land administration in northern Ghana.[13]

---

[12] The 1992 constitution of the Republic of Ghana divested and returned all lands in northern Ghana to the communities and it is being administered in accordance with customary law. Prior to the divestiture, authority over all lands in northern Ghana (before Ghana attained independence from Britain) were vested in the colonial governor for the use and benefit of the indigenous people. When Ghana became a republic in 1960, the constitution vested all lands in northern Ghana in the President for and on behalf of the people of Ghana.

[13] The structure of land tenure in Northern Ghana is very diverse with different opinions on who holds allodial rights to land in the area (see for instance, Benning, 1996; Kasanga, 1994). Among some clans, especially those with centralized chieftaincy systems (Dagbon, Gonja, Mamprugu, etc), the paramount chiefs hold the allodial title to lands in the respective

Generally, the customary laws confer greater recognition to group rights. The tindana grants usufruct rights to families or households. Each family to whom land has been allocated has the prior right to cultivate the land *in perpetuo*. However, ownership rights continue to be vested in the community, thereby restricting the family's right to dispose (sell, lease, mortgage, etc.) of the land. Nevertheless, each family's access to land is secure. Usufructuary rights can be bequeathed to, usually, male descendants of a family. This system has, through history, made male members of the communities obtain land and have secure land tenure. Customarily, women are not allowed to own land. Women who want to farm obtain land from their relatives (parents and brothers) and husbands. Women can, however, temporarily own lands of their deceased husbands (household heads) while they take over the responsibility of the household when their male children are too young to assume that role. Settlers can permanently have usufructuary rights to land as a gift from the first occupiers or their descendants, and thereafter possess similar rights to property as the original landowners. Strangers to a community can obtain usufructuary rights to land from the chief, tindana or landlords.

Any member of the community with the permission of the tindana can clear virgin lands, where they exist, for cultivation. No land rent is paid though some presents such as cola nuts, guinea fowls or a portion of the annual harvest are sometimes given to the tindana for sacrifices to the ancestors. The one who first cleared the virgin land "owns" it, although ownership does not give the right to sell or lease the land. Such ownership rights, as stated earlier, are usually customary rather than formal (Dittoh, 2000).

Land disputes occur often between different tribes over boundaries between tribal lands, between indigenous people and settlers along agricultural frontiers, between different authorities (chief and tindana) within the same tribe, and when leased land is wanted back by the original owners.

Fallow lands, virgin lands, forests and all land not allocated are regarded as communal property. Usable resources on the land may also be used by any member of the community. For instance, grasslands can be used as grazing fields, forest resources can be exploited for fruits, wood, meat and vegetables; rivers and ponds may be used for minor irrigation etc. However, there is an exception to the extraction of the fruits of economic trees (especially, dawadawa-*Parkia filocoides*, and shea nut tree - *Butryopernum parkii*). Though growing in the wild, only members of the household on whose land these economic trees are found can have the right to pick the fruits. It is ironic that even someone to whom a household has granted the usufruct right of land has no right over the fruit trees that grow on it. It is said that the fellow could even be

---

traditional areas. Under these systems land control is exercised through a hierarchy of divisional, sub-divisional and local chiefs in association with tindanas. However, among other tribes (Talensi, Kusasi, Dagaaba, etc.) allodial right is vested in the tindana. So depending on the tribe, land is administered by the village chief or tindana to individual compounds/families which in turn subdivide it amongst their members.

ejected from the land if he deliberately causes harm to any economic tree that grows on that land.

Although land tenure in the study is said to be communal, control over land is largely in the hands of family/clan units due to long occupancy and usage. The chiefs and the tindana have little say in the allocation of these lands, except in the areas where there are traditional shrines and groves, although they remain the final authority over the land. The same can be said of irrigation lands. In areas where the users of the irrigation facilities have total control over the command area, tenure is secure as long as those who have been given the usufruct rights demonstrate use by effectively cultivating their plots. In some schemes, however, usufruct rights are seasonal as the lands revert back to their original landowning households in the wet seasons (section 3.5.3).

2.6.4 Customary water rights

Under the traditional customary law, water is viewed as common property, managed for the common good of the community. Indeed, water resources (wells, streams, rivulets, rivers, lowlands) are a traditionally treasured natural resource, and held as a communal resource. Thus, water as visibly seen cannot be privately owned. Water sources, no matter on whose land they may be located, are free-for-all to members of the community and the general public at large. Chiefs, elders and priests (tindana) additionally control, manage and regulate the use of water sources (such as ponds, dugouts, dams, streams and wells). The use of water resources is shaped by historical convergence of environmental and political conditions that promote preservation of the resource, general access associated with use, multiple use and conflict avoidance. The traditional principles of communal ownership and ethnics of social control over water shape the principle of equity in water distribution that is central to the irrigation management reforms in northern Ghana (section 3.5).

As a common property, the community assumes responsibility for their common exploitation and management. Although the management of the irrigation schemes is entrusted to the users' associations, the communities regard the irrigation water as common property. Members who fail to turn up for communal work for the maintenance of water resources are sanctioned. Sanctions can include fines and even preventing one from using the resource for any purpose including livestock watering (Bakang and Garforth, 1998).

2.7   Lessons

This chapter examined the physical, political and socioeconomic setting of the study area. It also analyzed the role of traditional institutions in natural resource management in the study area.

Northern Ghana is characterized by low and erratic rainfall regimes, low soil fertility, rapid soil degradation, and high population densities in many parts of the region. We learnt that in spite of the harsh climatic conditions which make rainfed farming risky, agriculture remains the mainstay of the majority of the population of northern Ghana. Agricultural productivity is low. Poverty is widespread and food insecurity persistent in the region. Under the conditions of limiting ecological factors and a lack of opportunities for alternative income sources, households migrate to southern Ghana to seek economic opportunities. Efforts at boasting food security in the area led to the construction of several dams and dugouts for dry season gardening and livestock watering. Road network in the area is poorly developed and infrastructure for food markets very thin.

Household organizations and traditional labor exchange arrangements in the area promote the development of social networks, norms and trust upon which people tend to cooperate for mutual benefits. The patriarchal household arrangements make social mobilization for collective action easy.

Traditional institutions have long played important roles in natural resource management. Indigenous people's knowledge, practices, values and capabilities form an important basis for natural resource management in the local communities. Traditional institutions function to activate and enforce social norms of behavior through established patterns of authority and leadership.

Property rights systems that govern access, use and management rights of natural resources in the traditional systems emanate from the communal systems in which households have usufruct rights to the resource components.

In the context of irrigation management, responsibility for good performance is supported not simply by bureaucratic rules but also by norms and social sanctions. Traditional institutions impact efficiency through their influence on resource allocation and the sets of norms and exchange relationships the agents must assume. Social contracts in the indigenous societies also serve to re-enforce the maintenance of a productive resource from generation to generation. Individuals in the local communities are constrained by ethical and moral values that induce cooperative behavior.

Adherence to traditional customs, religions, norms and taboos dominate the lifestyle of the people in the rural communities of the study area. However, the changing perceptions about traditional believes and the spirituality (of natural resources), and the declining authority and functions of the Tindana remain a challenge to the ability of traditional institutions to effectively enforce resource use regulations.

The final sections of this chapter have attempted to bring to the fore the fact that in a given society, culture, institutional arrangements, and attributes of people's behavior affect the outcome of collective efforts. Most of these attributes are, however, difficult to change in the short run. Therefore, improving the capacity of the traditional systems to develop systems of cooperation and effective systems of monitoring and enforcement of community regulations is the key to the success of natural resource management strategies at the community level.

3   EVOLUTION OF PARTICIPATORY IRRIGATION MANAGEMENT PRACTICES IN NORTHERN GHANA

3.1   Introduction

As a response to the need to meet the food needs of the growing population of northern Ghana under conditions of low and erratic rainfalls and low factor productivity in rainfed agriculture, irrigation development has received significant attention in Ghana. Several small dams have been constructed to conserve water for livestock watering and dry season gardening. However, state management of small scale irrigation schemes in particular has been fraught with problems. As a result of inefficiencies associated with state management, participatory irrigation management and irrigation management transfer are management concepts being actively promoted in the region in recent times. This chapter examines the dynamics of irrigation management in northern Ghana with the view to understanding the impact of institutional and organizational efforts on the efficient management of irrigation schemes. It also highlights some common problems and conflicts. The chapter begins with background information on irrigation development in Northern Ghana.

3.2   Irrigation development in northern Ghana: An overview

The potential of agriculture in northern Ghana, with its large expanse of land and many river basins, has long been realized by successive Ghanaian governments since independence (see Konings, 1996). However, the unreliable weather conditions, in particular the prevailing low and erratic rainfall pattern were considered as a major obstacle to agricultural development in the northern part of Ghana. Therefore, the need for irrigation schemes to reduce risk in rain fed farming in the area was recognized. Several irrigation schemes were proposed to facilitate the production of grain and cash crops to raise the standards of living of the people and to turn northern Ghana into one of the largest grain baskets of the nation.[14]

Traditional or indigenous irrigation systems, constructed with local technology, controlled and managed by local people in response to their felt needs, have been in practice in most parts of northern Ghana since the pre-colonial period (Ayariga, 1992). As Underhill (1990) observes, these systems of irrigation have been practiced since time immemorial in different forms according to local conditions. In the Dagbon area in the Northern Region, residual moisture from lowlands is used to irrigate dry season vegetables while

---

[14] See 1. Ghana: Seven-Year Plan for National Reconstruction and Development. Financial Years 1963/64-1969/70, Accra: GP 1964 pp 61 & 64; and Ghana: Five-Year Development Plan, 1975/76-1979/80, Accra: Ministry of Planning, 1977. Pt II, page 118

water from shallow wells often dug in the river beds are used for irrigation in the Upper East and Upper West regions.

Formal irrigation systems were first introduced in northern Ghana (particularly, Upper East Region) in the early 1950s under the then Land Planning Unit, which later became known as the Irrigation, Reclamation and Drainage Division of the Department of Agriculture in the pre-independence era. Following Ghana's independence, the government had an ambitious program to transform northern Ghana into the country's bread basket and irrigation development became one of the features of the development vision of that time (Konings, 1986).

The pace of irrigation development in Ghana was, however, raised in the 1970s when it was realized that the policy of national self-sufficiency in crops and livestock production could only be achieved with the implementation of vigorous programs for the development of irrigation facilities.[15] The Ghana Irrigation Development Authority (GIDA) was established in 1977, as an autonomous institution within the then Ministry of Agriculture, with the mandate to perform all irrigation related activities in Ghana.[16] In the pursuance of this policy, irrigation schemes of varying sizes were developed to provide water for large livestock populations and for vegetable gardening in the dry season. The Ghanaian government, with the support of the World Bank and IFAD, instituted programs and projects in northern Ghana, including the Northern Region Rural Integrated Program (NORRIP) and the Upper Region Agricultural Development Program (URADEP) in the past, and recently the Upper East Land Conservation and Smallholder Rehabilitation Project (LACOSREP) and the Upper West Agricultural Development Program (UWADEP), which have and continue to make a significant contribution to irrigation development in the study area. Several NGOs and other government-funded projects in northern Ghana (e.g., Village Infrastructure Project (VIP)) are assisting to construct more dams in the study area for dry season farming and livestock watering. Nevertheless, the area under irrigation in the country remains very limited (GPRSP, 2002)[17] and the use of ground water from hand-dug wells is still common for irrigation garden plots in the area.

Generally, four types of irrigations systems can be identified in the area: (i) large scale irrigation systems (using large dams and large networks of canals, laterals and sub-laterals); (ii) small-scale irrigation systems (using small dams, consisting of small irrigable area served by canals from the dam, and dugouts); (iii) small-scale pump systems (using motorized pumps to draw water from rivers and streams); and (iv) traditional "bucket and calabash" systems (using

---

[15] Ghana: Five-Year Development Plan, 1975/76-1979/80, Accra: Ministry of Planning, 1977. Pt II, page 118

[16] IDA was established under the Supreme Military Council Decree (SMCD) 85 of 1977, with its headquarters at Zuarungu near Bolgatanga in the UER.

[17] Ghana Poverty Reduction Strategy 2002-2004. An Agenda for Growth and Prosperity. Analysis and Policy Statement. Accra, Feb. 20, 2002.

wells, streams, ponds, and other water bodies) (see also Dittoh, 1998). However, the dominant systems in northern Ghana are the first two, with the small-scale (dam and dugout) being the most prominent. In fact, there is a growing trend in Ghana towards the construction of small-scale farmer-managed systems due to the high costs of construction and management (including technical expertise required for the operation and maintenance) of large scale irrigation systems.

The dam-based irrigation systems in the study area are of the simple gravity type, based on surface flow of water from dams fitted with inlet and outlet valves. Reservoirs are designed and located to harvest water along water courses from a defined area referred to as the catchment area, and an irrigation area down stream of the reservoir located beyond the dam embankment. This area is watered through network of canals from service valves, controlling flows of water to the irrigation area. Spillways are constructed to carry away excess water in bid to safe the embankment in periods in which the water quantity is beyond the carrying capacity of the reservoir. Typically, many of the small-scale irrigation schemes in the study area have two main canals (left and right canals) enclosing the irrigable area, which are branched into several laterals then sublaterals which direct water to farmers' fields. There is also a main drain at the bottom of the valley that drains out excess water. Many of the canals are made of concrete slabs while the laterals and drains are embanked by mud.

The small-scale pumps and traditional systems are completely owned, controlled and managed by the individual irrigators, who take management decisions and bear their own risks (Dittoh, 1998). In northern Ghana, the traditional irrigation systems involve fencing out individual plots with either thorny bushes/shrubs or mud walls along valley bottoms, seasonal rivers and ponds. Shallow wells are constructed manually in dried river beds by farmers each year from which water is drawn either with motorized pumps or with buckets, gourds or calabashes for vegetable gardening in the dry season. This form of irrigation is often more labor intensive than the formal irrigation systems. There is rapid growth of pump irrigation especially along the Red and While Volta, where motorized pumps are used to draw water from the rivers, and also excess water from irrigation drains.

Close to 300 dams and dugouts[18] were developed for irrigation and livestock watering throughout northern Ghana. However, many of the schemes have collapsed. Figure 3.1 presents the distribution of the existing dams and dugouts in the study area. The majority of these small dams were built by GIDA in the 1950s and 1960s. The control, operation and management of the larger schemes were largely under the GIDA. Indeed, apart from private/NGO schemes all irrigation systems in northern Ghana were in the past operated and maintained by the Ghana Irrigation Development Authority and its subsidiaries.

---

[18] The distinction between a dam and a dugout as applied in this work is that while a dam has an outlet for releasing water for irrigation downstream, a dugout has not and irrigation is done by harvesting water from the reservoir. Dugouts are often too small for irrigation and are normally intended for livestock watering and domestic use.

Figure 3.1: Distribution of dams and dugouts in northern Ghana.

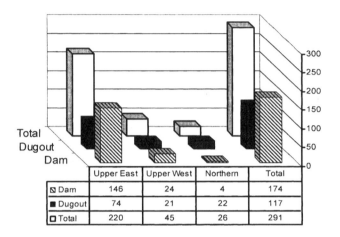

	Upper East	Upper West	Northern	Total
▨ Dam	146	24	4	174
▪ Dugout	74	21	22	117
▢ Total	220	45	26	291

The total irrigable areas under the smaller schemes range from 4ha to 55ha. Thus, per capita land availability is quite small, averaging 0.13ha (Engel, 2003).

3.3   Management of formal irrigation schemes in northern Ghana

It was the expectation that irrigated agriculture would become so profitable that the economic activities it would generate pay at least for the operation and maintenance costs. However, emerging consensus among irrigation experts and financing agencies is that irrigation is not performing anywhere near its potential (Hotes, 1982; FAO, 1987). Public irrigation schemes have rather become an economic burden to governments in the developing world and Ghana is not an exception. Evaluations of public irrigation systems have shown that in most of the schemes services have deteriorated due to faulty design and construction, as well as neglected maintenance and inefficient operations. This section examines past and present management interventions in the irrigation sub-sector as well as policies intended to streamline planning, development, operation and management to reduce costs and improve efficiency. Emphasis here is on the formal large- and small-scale irrigation systems.

3.3.1 Large and medium scale schemes

The larger irrigations schemes in northern Ghana are the Tono and Vea schemes in the Upper East Region, and Bontaga, Golinga and Libga in the Northern Region.

Tono and Vea, the two largest irrigation schemes in Ghana, are managed by the Irrigation Company of Upper Region (ICOUR), a semi-autonomous government agency, which is expected to increasingly commercialize to become self-financing (Dittoh, 1998). The Ghana Irrigation Development Authority manages the Botanga and the Golinga irrigation schemes. A top-down approach in irrigation management initially adopted by these organizations resulted in management problems, institutional weaknesses, and a lack of cooperation from farmers (Dittoh, 1998). The dominant role of the state in the management of the schemes made farmers more dependent, and bred a psychology of paternalism which did not encourage local assumption of or participation in maintenance responsibilities. In particular, GIDA-managed schemes suffered the most leading to the fast deterioration and collapse of many of the dams. Indeed, the irrigation sub-sector review in 1986 (World Bank, 1986) revealed that the large-scale irrigation projects in Ghana had a poor record of success. The schemes were beset with high operation and maintenance costs, a low level of community participation in operation and maintenance, and the deterioration of the physical irrigation infrastructure due to a lack of funds and maintenance. Farmer organizations were not institutionalized and in many cases farmers only came together, at the scheme level, to afford them the opportunity to obtain group loans and credit from banks. Farmers hardly had any roles to perform apart from tending their plots and, until recently, did not even pay for irrigation services (Agodzo et al, 1998).

The organizational gap between farmers and irrigation managers led to apathy and inefficient use of water. At the heart of it were a lack of interest on the part of the farmers to participate in the maintenance of the schemes, a lack of concern for the success of the projects, persistence in the expectation that project authorities should maintain the system on which their livelihoods depend, conflicts between project management and farmers, and deliberate abuse of the system through vandalism and illegal tapping of water. As the irrigation projects expanded, it became obvious that government could not afford to maintain the free cost irrigation objective it heralded as a means of encouraging irrigation agriculture to boast agricultural production. Lack of funds for maintenance of the systems resulting from reduced budgetary allocation to the irrigation subsector led to neglect, deterioration and the subsequent collapse of many of the schemes. Indeed, GIDA's active rule in irrigation management was cited in a World Bank (1986) review of Ghana's irrigation subsector as one of the major causes of poor performance of a majority of the irrigation projects. Following the recommendations of the World Bank report, attempts were made to encourage farmer participation in some of GIDA's schemes (Sam-Amoah and Gowing, 2001).

The need to collaborate with the beneficiary communities to organize operation and maintenance has been recognized and farmer participation has been given a greater role in all the GIDA managed schemes. It is believed that increased involvement of local communities and the transfer of management responsibilities of the schemes to beneficiary communities will facilitate internalization of externalities associated with the use of resources on the schemes, and enable users to cooperate and coordinate the use of the dam resource (see for instance, Vermillion, 1997; Tang, 1992; Meizen-Dick et al, 2002; Kamara et al, 2001).

ICOUR has adopted a participatory irrigation management strategy, which aims at increasing the role of participating farmers in the operation and management of the schemes. Communities under ICOUR schemes have been organized into irrigators' associations, referred to as Village Committees (VCs). The roles of the village committees include land allocation at the community/village level, irrigation water control and management, as well as maintenance of secondary canals, field drains and field bunds. VCs are also actively involved in the protection of the catchment area and the settlement of disputes. At the project level, a land allocation committee, headed by the District Chief Executive, allocates land to VCs, who then sign tenancy agreements with ICOUR stipulating, among other things, that maintenance activities are to be undertaken by the VCs. The village committees in turn share their allocated land among members. A consultative committee, consisting of farmer representatives, the district administration and ICOUR management, meet before the beginning of each crop season to fix water levies. ICOUR provides guidance to promote equity and fairness, allocates excess land to contract farmers, and collects water charges. The participation of farmers not only makes it easy for ICOUR to deal collectively with the farmers but also for the company to perform its managerial activities smoothly. This strategy has improved communication between farmers and the company and has reduced conflicts. Similarly, GIDA has been encouraging greater farmer participation in the management of its schemes at Botanga and Golinga, near Tamale. Farmer associations have been formed by the irrigators, with their roles and responsibilities being similar to those of the VCs. In practice, however, responsibilities are assigned to local committees without accompanying rights such that they become little more than an extension of GIDA. Ownership and operational responsibilities for the irrigation system largely remain under the control of the irrigation development authority.

Observations from the participatory management practices in GIDA schemes show persistent difficulties in getting farmers to cooperate, especially in scheme maintenance. Even in ICOUR some farmers see themselves as only purchasing irrigation and mechanized services from the company and thus expect the company to undertake all maintenance tasks to improve upon their services. In other instances reluctant farmers often see their group leaders as performing some of the roles of GIDA (Agodzo et al., 1998). Thus, getting farmers to participate in operation and maintenance still remains a difficult and complex task for GIDA as beneficiary communities perceive these dams as state property and expect the government to maintain them.

3.3.2 Small-scale schemes

Many small-scale irrigation schemes based on earth dams and dugouts were constructed throughout northern Ghana between 1950 and 1965. Quite a number were also funded under World Bank projects including the Upper Region Agricultural Development Project (URADEP) in the 1970s. GIDA managed all the schemes until URADEP took over the management of the small dams in the Upper Region in 1976. GIDA's technical officers stationed at the schemes combined the roles of construction foremen, agronomists and *de facto* managers. They were in charge of land allocation, operation and maintenance of the schemes with funds from the state (see Ayariga, 1992). Under the URADEP, the Small-Scale Irrigation Division (SSID) was set up to take over the responsibility for operation, maintenance, rehabilitation and construction of small dams. The SSID management strategy, however, did not differ much from that of GIDA. Rights and responsibilities were not clearly defined and technical officers were largely in charge of the operation and maintenance of the schemes. This did not only promote dependency but farmers also felt alienated and apathetic towards maintenance of the schemes. The irrigation schemes were seen as government property and their maintenance as the government's responsibility. Lack of state funding and poor maintenance (in many cases complete lack of it) resulted in deterioration and in a state of disrepair of many of the schemes.

Nevertheless, the value of small scale irrigation, especially in northern Ghana where rainfed agriculture cannot meet the demands for basic food due to droughts and floods that often result in total crop failure, cannot be overemphasized. These experiences certainly affirm the need for the rehabilitation of the irrigation systems that have fallen into disrepair and perhaps the construction of new ones. Similarly, rehabilitation cannot be limited only to physical infrastructure; management systems are generally just as badly in need of overhaul as the irrigation infrastructure (FAO, 1987).

Several donor agencies, government organizations (e.g. MoFA, GIDA and Village Infrastructure Project of MoFA, etc.), and NGOs are involved in the rehabilitation of these schemes and the construction of new ones, which are to be managed by farmers (Dittoh, 1998). The major rehabilitation schemes in Northern Ghana are the IFAD-funded Land Conservation and Smallholder Rehabilitation Project (LACOSREP) and Upper West Agricultural Development Project (UWADEP). In its first phase, LACOSREP rehabilitated a total of 44 dams and dugouts (IFAD, 1999), whilst 14 dams were targeted under the

UWADEP. Phase II of LACOSREP, which runs until 2005, envisages the rehabilitation of 30 old dams and the construction of 12 new dams. World Vision International, the Red Cross, and the Catholic Church are among the NGOs prominent in irrigation development in northern Ghana. IFAD and other donor organizations involved in irrigation development continue to spearhead the promotion of beneficiary involvement in planning and management as essential for the sustainability of the irrigation systems.

3.4 Participatory irrigation management: Earlier experience in northern Ghana

Experiences with community participation in irrigation management in northern Ghana started during the early 1970s when, under the URADEP, the small-scale irrigation division was created to assume the responsibility for the construction of small scale irrigation systems, with the participation of farmers in the management of the schemes.

Under the URADEP concept, a technical officer supervised all SSID dams in each district while farmer committees were set up to take over the responsibilities for the operation and maintenance of the schemes. Indeed, community participation was advocated in the operation and maintenance of the schemes, but seldomly practiced in reality. For the most part, these committees largely existed in name, and where they were found to be functional, the operation of the schemes was in many cases carried out by the SSID technical officer (Ayariga, 1992). Farmers' roles in management and maintenance responsibilities were not clearly defined.

Supervision was weak and the staff lacked experience in managing irrigation schemes which resulted in poor management (ibid). Most of the small dams have broken down due to general neglect of the dams and farming in the catchment area (Ayariga, 2002).

Ironically, the SSID staff lacked the motivation to organize farmer participation as many perceived farmer participation in management to be an impediment to rent seeking opportunities. There was no motivation to empower the farmers' committees with real authority, land allocation, water distribution and maintenance responsibilities. Farmers were made to play passive roles in the management of the schemes, and therefore lacked a sense of belonging, ownership and responsibility to carry out operation and maintenance activities. But as long as the immediate incentives (i.e., the "free" inputs and credit)[19] kept flowing, the farmers' committees survived to transcend deep tribal, political and social differences, which resurfaced yet before URADEP folded up (Ayariga, 1992). General apathy on the part of the users led to fast deterioration and collapse of many of the schemes. The conclusion made by the World Bank in

---

[19] High default rates were recorded in the recovery of loans and input credit (see Ayariga, 1992)

the irrigation sub-sector review (1986) that the project failed to establish successful farmer management of small-scale irrigation systems was therefore not startling.

3.5 The WUA concept and methodology in northern Ghana

3.5.1 Motivation

There is no doubt that experiences from the failures of past irrigation management strategies and other agricultural development initiatives in the study area have greatly informed current interventions under LACOSREP and UWADEP as well as the general policy of resource management transfer to local institutions currently taking place in Ghana as a whole. A core component of Ghana's small scale irrigation development policy reform is to turn over the operation and maintenance responsibilities as well as authority and control of the irrigation systems to the beneficiary communities (Engel and Gyasi, 2002). It is believed that operation and maintenance will be improved by maximizing the collective participation of farmers. This section dwells on the methodology for forming Water Users' Associations (WUAs)[20] under the LACOSREP as it is this institutional arrangement that is being replicated to many other projects across the study area.

The LACOSREP strategy has been to get beneficiary groups involved in all stages of rehabilitation and construction with the aim of turning over the responsibility of operation and maintenance to the user groups after completion. The intention is to strengthen the sense of belonging, ownership and moral responsibility of the beneficiary communities to carry out better quality operation and maintenance, which is necessary for the long term sustainability of the facilities. Retrofitting dams with user groups, as had previously been the case, had made the attainment of project objectives more difficult (IFAD, 2003).

In the beginning (LACOSREP I and early stages of UWADEP), WUAs were formed at the same time as the dams were being rehabilitated. In some instances, it turned out that some farmers were motivated to participate mostly by the chance to earn cash income (through payment of some minimum wage and food rations) but not in the view to subsequently become members of the WUA. Indeed, the strategy at the time put more emphasis on the physical construction works, without adequate emphasis on institutional aspects essential to sustain and improve performance.

However, LACOSREP II deviates from the past by seeking to promote farmer participation at the pre-construction, construction and post-construction stages of the project. Apart from making the intervention demand-driven, this

---

[20] FAO (1982:8) defines it as an organization of water users that manages, allocates and distributes water from a common source in the most efficient participative manner to benefit all the members

strategy also seeks to improve farmer participation in the planning and implementation phases of the rehabilitation program. User groups are being formed and getting registered as cooperative organizations as a means of extracting significant organizational contributions upfront, before rehabilitation starts. It is envisaged that this approach will have positive effects on the beneficiaries' attitude towards operation and maintenance of the schemes. Relevant government institutions and NGOs conduct initial animation prior to detailed survey and design work by the irrigation development authority. Labor intensive rehabilitation techniques have become a major opportunity to orient the farmers to maintenance techniques even before rehabilitation is completed, and to strengthen the sense of ownership and moral commitment of the user group to ensure quality maintenance of the facilities.

3.5.2 The Structure and organization of the WUA

Typically, the local unit of organization under the participatory irrigation management is the water users' association. There has not been any universal structure, but a typical WUA in the study area is an umbrella organization that encompasses affiliated groups in the beneficiary community that have stakes in the dam infrastructure (as means of livelihood) and organize around economic interests. The dominant economic interest groups here include irrigators (gardeners), livestock owners, and fishermen who organize themselves into associations. This does not limit the right of any member of the beneficiary community from becoming a member, however. This structure of the association has been promoted by the LACOSREP with the recognition that there could be conflicts over the use of the dam water. It is thought, for example, that gardeners would like to expand to irrigate a lager area beyond the designed capacity of the reservoir so as to admit more people. This may be to the detriment of fishermen and livestock as eventual water shortages might affect their interests. Fishermen could also use pollutants to harvest fish which could be injurious to the economic interests of the other groups. The coming together of these interest groups could therefore help to promote mutual interests and foster harmony. Although each individual maintains control of his/her economic activities at the scheme, the group acts cooperatively to manage the system. Figure 3.2 below depicts the structure and responsibilities of a typical water user association under LACOSREP.

Elected representatives of these interest groups constitute the executive committee of the WUA, also referred to as the Damsite Management Committee (DMC). The DMC is the overall management body for the damsite and has the responsibility to enforce all bye-laws enacted by the association. Where several villages are sharing the use of an irrigation facility, it is common to find, primary sub-groups and their elected representatives serving on the DMC. Members of the executive committee, consisting of the chairperson, secretary, treasurer, and other members, are elected by all members either by consensus or secret balloting in a general assembly. The size of the DMC membership varies across schemes and depends on the size of the WUA, but elected officials serve for specific periods ranging from 2-3 years. Typically, executive committees meet fortnightly. The executive committee members are not paid, and their services (time contributions) are strictly voluntary. Major decisions of the association are taken in meetings of all members, although the executive committee can take routine decisions. Officials of relevant public institutions and NGOs are encouraged to attend meetings of the WUA, by invitation, to offer technical guidance.

Figure 3.2: The structure and functions of water users' association in northern Ghana.

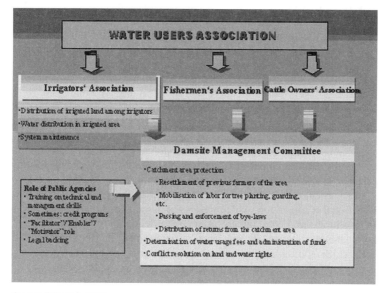

Source: Edig et al, 2002

The main functions of the user groups include the overall responsibility for the operation, administration and maintenance of the irrigation infrastructure, land allocation, water distribution, fixing and collection of water fees, and designing and enforcing bylaws. The WUAs also resolve disputes and conflicts among members. GIDA is expected to provide supervision and assist the farmers to take care of the maintenance of head works, primary canals and other major structures. Yet no observation was made of a program for annual (joint) inspection of the schemes to provide technical advice, promote preventive maintenance and to ensure that potentially expensive and dangerous problems beyond the capacity of farmers are identified and dealt with on time. With assistance from the extension service and IDA, farmers have also been involved in the planting of grasses on the embankments and protected belts to prevent erosion and reduce siltation of the dam reservoir, although with little success (Ayariga, 2002).

The damsite management committee ensures that the WUA carries out its functions. Indeed, the WUAs are responsible for designing their own bylaws although in some cases prototypes were presented to them by the Project Coordination Unit (PCU) of the LACOSREP for modification to suit local conditions before adoption. The bylaws encompass the entitlements (Vermillion, 1994) and obligations of the water users in the schemes. Rules and sanctions to discourage deviant actions are also spelt out in the bylaws.

In both LACOSREP and non-LACOSREP dams, training of WUA executives is provided free of charge by the PCU in coordination with the District Project Management Unit (DPMU). To ensure continuity in case of high turnover within the executive committees, a number of literate members are trained. Training topics cover canal lining, routine maintenance, water allocation and water distribution systems, irrigation agronomy, and organization and management as well as cooperative norms. Training is conducted by the extension services staff of MoFA, GIDA, and the Department of Cooperatives and Community Development. Other support comes in the form of technical expertise, especially on the infrastructural aspects of the dam (Ayariga, 1993).

In many of the associations, a well-defined leadership structure has emerged and there are committees such as land allocation, water distribution, maintenance, finance/levy, and disciplinary committees.

The structure of the WUA as presented in Fig. 3.2 depicts a multifunctional organization, which makes the associations more likely to build up both resources and commitment to operate effectively over time. The set of functions are not universal but usually determined by the organizational structure which is often shaped according to the goals of the WUA. However, distributive and conflict resolution functions appear to cut across systems with the objective of ensuring equity and social harmony in the irrigation communities. Generally, labor mobilization, work planning and distribution of work among members are organized and coordinated by the damsite management committee, and in some WUAs, by subcommittees responsible for maintenance.

Equitable distribution of irrigable plots among gardeners, for instance, is being stressed in all the small-scale irrigation communities in the study area in accordance with the broad objective of the provider organizations to alleviate poverty. The WUAs are being encouraged to evolve plot allocation mechanisms that ensure equity. The executive committees of the WUAs are responsible for the allocation of plots. Sometimes they do so through a sub-committee often referred to as the land allocation committee.

3.6 Land allocation

As can be deduced from section 2.6.3, land constituting the irrigation area is owned and largely controlled by individual households as the result of long-term usage. To ensure that the economic benefits of the schemes are shared by all, the beneficiary communities were made to agree to land redistribution[21] that ensures that the poor, especially women and landless members of the community, gain access to irrigable land for gardening. Also, households who in the process of the dam rehabilitation were displaced from the area that now constitute the reservoir and/or the protected catchment area were to be resettled in the irrigation area. Catchment area protection, which is important for preventing the siltation of the reservoir, will be difficult to achieve if the displaced farmers are not resettled in the irrigation area (Ayariga, 1995).

No compensation was offered to the original landowners in return for their land. However, attempts are being made to get District Assemblies to legally acquire the scheme areas, sign user agreements with the WUAs and use part of the user fees to be paid by the WUAs as compensation for the landlords.

By design, land allocation was to be the task of a WUA executive or subcommittees mandated by the association for this task. The expectation was that any allocative principle adopted by the WUA would be transparent and lead to equitable allocation of land. However, it, becomes apparent that due to high population density and the resultant land scarcity that characterizes most parts of the study area, it is difficult to insist on a WUA taking total control over irrigation land when landowners are not adequately compensated for their land. Perhaps the most credible option was for a WUA to enter into agreements with the landowners that would give it some control over the land resources to ensure secure land rights even if temporary (Nyari, 2002).

In spite of the promises by land owners and community leaders to hand over the irrigation land to the WUAs for redistribution, some landowners continue to either resist or interfere with the land redistribution policy. In 38.5% of the communities surveyed, the original landlords interfere in the allocation of land. In particular, the temptation for landowners to either interfere in land allocation or resist land redistribution, all together, is high in communities where land is very scarce. In most of the land-scarce communities, the WUAs actually bemoan their inability to compel landlords to give up their land. WUAs thus have little control over land allocation in such communities. It was observed that in some cases the landowners who are not interested in gardening preferred to have tenure agreements (including share cropping) with prospective gardeners in an attempt to maintain total control of their lands.

---

[21] In actual fact, willingness of the land owners to share their land with other members of their communities was one of the criteria for selecting schemes for rehabilitation. The communities were made to understand that it is not acceptable for large sums of money to be used to rehabilitate schemes for the benefit of only a few individuals.

In many other communities, however, we observe a shared interest in the dam, a situation that allows communal interest to override individual interest. This communal attitude has been manifested in the readiness of many landowners to share land with other community members. Whilst in some communities irrigation lands have readily been handed over to the WUA for redistribution, in others the communal land tenure systems have been flexible enough to allow the emergence of new land ownership forms that accommodate community interests (see also Nyari, 2002). Indeed, compromises exist in many of the schemes in the form of dual control over land, in which the original landowners use the land during the rainy season for main season farming while the irrigators' association takes over the land in the dry season. This arrangement has helped to hold in check potential land conflicts.

In allocating land, the WUAs tend to give priority to farmers displaced from the catchment area to discourage them from farming within the catchment area. In most of the schemes consideration is given to every resident in the community interested in gardening. Indigenes constitute the bulk of the beneficiary group in six (11.5%) of the surveyed schemes.

Land is allocated to individuals. Thus, some households may have more than one member with irrigable plots, but the principle is that at least one member of every household in the community interested in dry season gardening should be allocated a plot. Plots are not reallocated until the member abandons it or has the plot taken away from him or her for noncompliance with WUA regulations.

3.7   Conflicts and sources of disputes

It has long been argued that conflicts and disputes could be useful in human societies or organizations as a means of providing creative impetus to continually learn about problems, change relationships and social structures to eliminate causes of alienation, dissociation and antagonism and to develop new unifying norms (See Groothis and Miller, 1994; Mangorian, 1992; Coser, 1956). However, it becomes troubling where conflicts become rampant and destructive to life and property.

In northern Ghana, conflicts are gradually becoming a common feature. Many of the conflicts arise out of chieftaincy and landownership disputes (Ayariga, 1992) and often do not help the course of development. Land owners whose consent may be needed for irrigation projects may be deeply involved in conflicts, exacerbating the feelings and tensions that fuel division (ibid). The conflictual relationships more often than not disturb social harmony, derail progress and hinder cooperation and collective decision making. Divisions along fault lines discourage decision-makers from trusting communities to work cooperatively, as they hinder all forms of progress that require participation of beneficiary communities. In spite of its negative effect on cooperation, conflicts have been a common feature in common pool resource management. When they do occur, conflicts are managed internally by the WUA, and in complex situations the intervention of the traditional authority is sought for the settlement of disputes (section 2.6.1).

At the scheme level, most of the conflicts arise mainly over land and water allocation (see section 6.4). In some cases, original owners of land in the scheme area resist voluntary sharing of plots with other community members. This often results in intra-community conflicts and also frustrates maintenance efforts. This is because regulations on maintenance and catchment area protection will be difficult to enforce if only few farmers have an interest in the scheme and the majority, including those displaced by the reservoir and the protected catchment area, is not given land in the irrigated area.

Lack of probity and accountability in some of the WUAs are disincentives to collective action for the management of the schemes and sometimes spark disputes between members and their executives. Embezzlement and misappropriation of WUA funds and the failure of some executives to organize democratic elections breed mistrust and apathy in WUAs. More so, refusal of past executives to hand over elected positions to in-coming ones has been a source of tension, reduced credibility, and increased reluctance of some WUA members to pay levies.

3.8 Lessons

This chapter reviewed irrigation development programs in northern Ghana dating back to Ghana's colonial past. Past and present irrigation management strategies as well as prevailing socioeconomic contingencies which could enhance or hinder farmers' participation in the management of irrigation schemes have been examined.

In reality, irrigation development has received attention in Ghana's effort to meet the food needs of the growing population of northern Ghana, where annual rainfall is low and erratic, but the potential for irrigation development is high. The important role irrigation plays in the livelihood strategies of the communities in the study area means that effective management of the schemes is very crucial. However, irrigation development in the past was largely considered as a technical process. Little consideration was given to farmers' participation in irrigation development and management. State management of the small scale schemes became fraught with difficulties and previous attempts at involving farmers in irrigation management failed due to design and implementation problems as well as a lack of motivation on the part of the implementation agencies.

In resonance to the world-wide policy of devolution of the management of natural resources to user groups, a turnover of the management of small-scale irrigation schemes to user groups can be observed in northern Ghana. Water users' associations have been formed to which the management of rehabilitated schemes has been entrusted. The structure of the WUAs in northern Ghana is, however, influenced by economic, institutional, social, political, and cultural factors which can affect the efficiency of the organizations and indeed the importance of the WUA, especially where traditional cultures are allowed to influence the institutions for managing the resource.

The WUAs are entrusted to lead the collective management by farmers of the irrigation systems for the sustenance of the livelihood of members. The executive committees oversee the planning, operation and tasks related to resource allocation, maintenance and conflict resolution for the realization of the objectives of its members. As observed in a recent study (IFAD, 2003), the WUAs as a forum for managing the irrigation schemes (dam-wall, reservoir and catchment area as well as command area) represent an inspiring innovation, a platform for discussing and catalyzing collective action.

To be strong and successful in the discharge of their obligations, the WUAs should have a democratic leadership and they should be legally protected against interference, particularly from landlords and local chiefs. A legal framework that affirms the ownership rights conferred on the community and recognizes the WUAs as the farmers' representatives and managers of the irrigation facilities could make bylaws of the user groups legally enforceable and enhance the effectiveness of the organizations. Once properly registered and put on a secure legal basis, the WUAs can indeed become a vehicle for promoting participatory approaches to problem solving and decision making (see also IFAD, 2003).

4   SURVEY SETTING AND DESCRIPTION OF THE USER GROUPS

4.1   Introduction

In the previous chapter, issues concerning irrigation development and management reform in northern Ghana were discussed. The organizational structure of the local unit of management of the schemes that resulted from the irrigation management turnover has also been described. This chapter describes sampling framework and description of the sampled users groups and individual households. In particular, sections 4.2 and 4.3 of this chapter presents the setup of the fieldwork, and the description of the dataset to provide some understanding of how collective action is organized in the irrigation communities and the processes of participation in irrigation management at the local level. Section 4.4 highlights the maintenance activities undertaken by the Water Users' Associations (WUAs). Levy collection performance is also examined in section 4.5 while the remaining sections analyze social organizations and the multiple uses of the irrigation water in the survey communities.

4.2   Sampling procedure and data

This study was undertaken in two regions of northern Ghana (Upper East and Upper West regions) where community based irrigation management programs have been implemented. The two regions were selected not only because they boast of larger proportion of small-scale irrigation schemes in Ghana but also because the regions have been a focus of IFAD supported community irrigation management programs being implemented in Ghana. Institutions leading the establishment of water users associations in these regions are the Upper East Region Land Conservation and Smallholder Rehabilitation Project and the Upper West Region Agricultural Development Project. The study covered 58 irrigation schemes, six of which were not irrigating at the time of the survey (2 due to conflicts among users and the remaining for lack of water). Thus, fifty-two of the schemes were included in the analysis; 45 in the Upper East Region and 7 in the Upper West region. The field work lasted for six months; from January 2003 to June 2003. Community coordinates, and dam locations were recorded using the Global. Figure 4.1 presents a map of the Upper East and Upper West regions of Ghana showing the locations of the community dams. The list of all the communities and their geographic location are presented in Appendix A.

Figure 4.1: Map of Upper East and Upper West regions showing location of community irrigations schemes

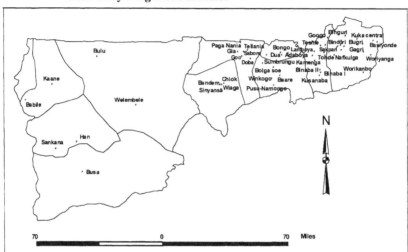

The survey was conducted in two rounds. The first round survey took us to all communities in the study area that have irrigation facilities, and involved technical evaluation of the state of the irrigation infrastructure to serve as the principal indicator of the user groups' success in mobilizing resources for the maintenance of the irrigation facility. With the assistance of a technical expert from the Ghana Irrigation Development Authority (GIDA) a checked list of irrigation system quality measures, adapted from a format used by IWMI to measure the impact of irrigation management transfer, were used to evaluate all the schemes. Close to 60 irrigation schemes were surveyed. However, schemes that had totally broken down and had not yet been rehabilitated were not included in the second round of the survey. This is because these schemes are not in use and there are no management structures in place.

The sampling technique was a two-stage clustered sampling in which we first purposefully selected communities with functioning irrigation schemes and where efforts have been made to institute water users' association as the management unit. The second stage involved random selection of households within each selected community.

Questionnaire pre-testing was conducted in two randomly selected communities. Response to the pre-test resulted in the revision of the questionnaire. Enumerators were recruited and a 2-day seminar was held to train the enumerators on the content of the sets of questionnaire (including concepts used) and on general enumeration skills. A field trial was conducted to check the enumerators' understanding of the questionnaire, after which a follow-up training was conducted and the questionnaire updated and finalized.

The main survey (second round) covered only communities with functional irrigation schemes. The survey was conducted in two levels: Community level and household level. Households were stratified into members of the irrigators association and non-members. The non-irrigating households included only those who did not have any member using the community dam to practise irrigation. Random samples of 10 and 6 were taken from members and non-members respectively. It therefore implies that, three sets of structured questionnaire were used in the survey: a community questionnaire and two sets of questionnaires for households. Table 4.1 presents the size and distribution of the sample.

Table 4.1: Community and household sample

Region	Community	Household		Total
		Member	Non-member	
Upper East	45	455	260	760
Upper West	7	70	42	119
Total	52	525	302	879

The community survey comprised eight modules. These are (i) social and demography, (ii) economic activities, (iii) history of irrigation, (iv) dam rehabilitation and uses, (v) scheme management, (vi) plot and water allocation, (vii) rule conformance, and (viii) scheme maintenance. The community questionnaire was divided into two sections. The section covering the socioeconomic characteristics of the community and history of irrigation in the community was administered to opinion leaders (Chief, Tindana, Assemblyperson, etc). The remaining section, which greatly concerned the user groups, was however administered to the executive members of the water users associations. The modules covered information on population, ethnicity, religion, community mobilization, social organizations, economic activities, market access, irrigation systems, WUA structure and membership, landholding, institutions and rule conformance, resource allocation, maintenance activities and contributions, levies, as well as training received by the WUA.

At the household level, information was gathered from the head of household with the aim of capturing household incentives to participate in collective action. A total of 521 households gardeners were interviewed; an average of 10 per community. About 300 non-irrigators were interviewed; mean of 5.8 non-member household per community. The household questionnaire consisted of comprised eight modules. The eight modules are (i) household demography, (ii) membership of local associations, (iii) household economic activities, (iv) irrigation activities (gardening), (v) land holdings, (vi) access to agricultural services (extension and credit), (vii) household wealth and assets, and (viii) household expenditure. The information from five members of the irrigators association were excluded from the analysis because those sets of questionnaires contained many unanswered questions. The analysis included 520 members of the irrigators' association out of the 525 members interviewed.

Prior to the fieldwork, an initial stakeholders' workshop was held in Bolgatanga in the Upper East Region in May 2002, which helped to define appropriate outcome measures and exchange of information on the state of community irrigation management in northern Ghana. The following sections describe the dataset.

Expert interviews were conducted to improve our understanding of the study area and help to generate local specific hypotheses. Data from secondary sources (mainly prices, crop production data, population, literature as well as project reports and documents relating to water and irrigation development in Ghana) were obtained to complement the survey data. Secondary data sources included Ministry of Food and Agriculture, Ghana Statistical service, Ghana Water Commission, Ghana Irrigation Development Agency and International Fund for Agricultural Development. A separate market price survey was conducted in four (2 urban and 2 semi-urban) markets to supplement price information obtained from the Ministry of Food and Agriculture.

4.3 Description of the local irrigation systems and user groups

As illustrated in the introductory chapters, the study area has the largest number of small-scale irrigation schemes, and has been central in irrigation management turnover programs in Ghana. Examples of user participation in irrigation management can also be found in the two major irrigation schemes in the region, Tono and Vea, which are managed by Irrigation Company of Upper Region (ICOUR). The focus of the analysis is, however, limited to the user-managed small-scale irrigation systems. In the course of the discussion emphasis is placed on socioeconomic factors that enhance group cohesion.

Water Users' Associations have been the effective tool for organizing farmers to take up management responsibilities. As illustrated in Section 3.5.2, a typical WUA consists of irrigators (gardeners), livestock owners and fishermen. The mean WUA size of the survey schemes is 231.5 with the smallest size of 15 at Posu Namongo, and the group at Gagbiri (with 928) accounting for the largest. Gardeners dominate the groups. Livestock owners and fishermen's associations were found in only 19.2% and 34.6%, respectively, of the surveyed communities (Table 4.2). In eight of the 52 communities both livestock owners and fishermen's associations exist along side the gardeners. In many cases the same members of the irrigators' associations show up as gardeners and fishermen, a situation that could make it difficult for the WUAs to assign to different groups different roles.

Table 4.2: Characteristics of the sample

Characteristics	Unit	Mean	Std. dev.	Mini.	Maximum
Characteristic of the irrigation system					
Year of rehabilitation	year	1997	2.29	1991	2001
Catchment area	Ha	190.41	215.30	30	1039
Reservoir capacity	1000m³	281.57	231.45	80	1170
Lined canals	m	1115.97	597.85	300	2900
Irrigation area	Ha	12.16	7.93	4	35
Rehab. was Labor intensive	(0,1)*	0.442	0.501	0	1
Irrigation extended by farmers	(0,1)	0.596	0.495	0	1
Government constructed	(0,1)	0.79	0.41	0	1
Characteristics of water users' association (WUA)					
Age of WUA	years	5.77	2.45	2	13
Size of WUA	no.	231	212	15	928
Gardeners assoc. members	no.	216	207	13	913
Women gardeners	%	37.27	15.80	5.81	76.19
Livestock owners association	no.	9.23	30	0	185
Fishermen's assoc. members	no.	6.48	11.33	0	50
Mean plot holding	ha	0.14	0.16	0.008	1.0
Gini coef. of plot sizes		0.35	0.13	0.03	0.72
Households per irrigation area	ha	23.18	21.42	1.17	127.53
Gross margin per unit water	¢/1000m³**	248,740	318,129	4,674	1,428,055
Gardeners before dam rehab.	no.	84.62	134.15	0	600
Community characteristics					
Population	no.	1817.37	1647.48	273	8351
Ethnicity	Index***	0.17	0.19	0	0.72
Distance to market	km	1.48	2.52	0	11
Villages using scheme	no.	5.31	2.9	2	15
Villages are of same ethnicity	(0,1)	0.75	.44	0	1
Irrigation experience	years	33.88	13.07	3	58

No. of observations = 52; * For dummy variables: 0=No, 1=Yes
** ¢ is Ghanaian currency (Cedis), US$1 = ¢8600 at the time of the survey
*** see below

The sizes (command areas) of the sampled schemes are very small ranging from 4-35 hectares with mean irrigation area of 12 ha (C.V = 0.65). The storage capacities of the reservoirs are generally small, and the amount of water available for gardening range from 80,000 - 1,170,000m³ (with the mean of 282,000m³), which the farmers complained have been decreasing due to silt deposition in the reservoir (see Appendix B for the method used in estimating water availability). The survey shows that the individual schemes in our sample are used by villages ranging in numbers of 2 to 15 with a mean of 5 per scheme. In close to 60% of the schemes, the communities have by themselves extended the irrigation area beyond the designed capacity of the reservoir in order to meet the demands for plots from community members. The implication has been frequent water shortages and a potential source of conflicts over land and water distribution.

In addition to the user-managed canal irrigation system there are several other forms of irrigation that are privately owned and operated in the study area, as discussed in chapter 3. These include shallow wells sunk to draw water for irrigation. The existence of private wells and other micro-irrigation systems, however, creates some degree of independence from the community systems and breeds apathy that can affect the maintenance of collective action on the community schemes. Although irrigators generally preferred to use the cheaper community-owned gravity irrigation systems, because it is much more costly (in terms of time and labor) for individual farmers to construct wells to irrigate than to participate in the maintenance of the communal systems, it appeared there is reduced willingness to contribute resources (labor) to the community-managed systems in the villages where irrigators have private systems.[22]

Women membership in the WUAs is generally encouraging, though there is much room for improvement. In at least 40% of the user groups, women constitute over 30% of number of gardeners.[23] This could also be the results of unintended positive effect of seasonal migration. This phenomenon can have positive impact on the poverty alleviation objectives of the projects setting up the schemes, given that women are the most vulnerable and high affected by poverty in the region (GSS, 2000). However, women have meager plot sizes not commensurate with their 30% share of the WUA membership.

Information about the ethnicity of the communities can also be important for understanding the underlying factors affecting the success of community resource management strategies, because patterns of cooperation and conflicts can be linked to social fragmentation. The study area has high ethnic diversity. In the Upper East Region, one can count several ethnic groups including Kusasi, Builsa, Frafra (Gurene), Kasena, Mamprusi, Nabdam, Talensi, Grushie, Busanga, Moshi, and other minor tribes. Similarly, in the Upper West Region one finds the Dagaaba, Wala, Sissala, Chakali, Lobi, Wangara, and Moshi among others. However, with some exceptions, the level of heterogeneity declines as one moves to the district and community levels, as depicted by the patterns of ethnic fragmentation indices presented in Figure 6.1 below. Following Alesina and La Ferrara (2000), we calculated the ethnic heterogeneity index (F_v) using the following relation

$$F_v = 1 - \sum s_{\kappa v}^2 \qquad \kappa = 1,....,K_v \qquad (4.1)$$

where s_{kv} is the share of respondents in village v who belong to tribe κ. There are K_v tribes in each village. The index F_v represents the probability that two individuals randomly drawn in the same village will belong to different tribes. The intra-district index (F_d) was constructed by replacing v by d (district). The mean village heterogeneity index is low (0.17); ranging from 0 to 0.72.

---

[22] Correlation coefficient for hours of labor for maintenance activities at the community scheme versus practice of other forms of irrigation is -0.089 (0.189; 2-tailed)

[23] We cannot rule out the potential of communities giving false information about numbers of women in their groups in order to please project authorities. Note that beneficiary communities are obliged to ensure that at least 30% of group members are women.

The left panel in Figure 4.2 depicts ethnic diversity across districts in the study area while the right panel shows the extent to which the villages are ethnically heterogonous (List of communities presented in Appendix A). As the figure depicts communities in the Nadowli, Jirapa/Lambusie and Lawra Districts of the Upper west regions are more homogenous, and are mostly inhabited by the Dagaaba ethnic group. In the Upper East Region, communities in Bongo and Builsa districts appear to be ethnically more homogenous. The heterogeneity of Bolgatanga, the regional capital, appear to overshadow the ethnically homogenous nature of the rural communities in the Bolgatanga District. However, communities in the Bawku East, Kasena-Nankena and Bawku West districts appear to be the most heterogeneous in the Upper East Region. A close look at the dataset revealed that most of the non-indigenes in the sample have lived in those communities for several decades (with mean of about 30 years of residence), suggesting possibility of cultural integration which could be useful for group cohesion and cooperation in collective action.

Figure 4.2: Ethnic diversity in the survey communities

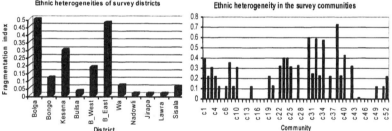

The potential of irrigation system to generate enough income to satisfy the income expectations of irrigators is a greater motivation for participation in collective maintenance of the schemes. The estimated gross margins per unit water are generally high across schemes (Appendix C). However, the high degree of variability (C.V=1.27) in the gross margin per unit water (as presented in Table 4.2 above) can have implications for the success of collective action for the management of the schemes, as this can affect the incentive structure.

Largely, water volume in the irrigation system is limited by the amount of water the dam can hold and of course the amount of rainfall received in the year. Volume also depends on the whether the dam is silted, strength of the dam wall to hold water, quality of the spillway, how well the catchment area has been protected and the quality of maintenance undertaken by the user group. Quality of maintenance is affected by the ability of the user group to make and enforce rules and by-laws, and resolve conflicts.

In the following sections we examine the maintenance activities undertaken by the WUA and what influences household decisions to participate in the maintenance activities.

4.4   Collective maintenance activities

Past experiences of deterioration and collapse of irrigation schemes in the study area, resulting from neglect make maintenance a core responsibility the WUAs are enjoined to undertake.[24] In this section we examine maintenance activities carried out by the user groups aimed at slowing down deterioration to sustain the performance capacities of the schemes. Typically, the most important maintenance activities carried out by user groups include: (i) maintenance of the irrigation system (comprising control structures, network or canals, laterals and drains), (ii) dam infrastructure (dam wall, spill way, and reservoir), and (iii) protection of the catchment area. Table 4.3 presents a list of maintenance activities carried out by the user groups across the survey schemes.

Table 4.3: List of maintenance activities

a. Canals, laterals and drains
   - Clearing main canal of weeds and silt
   - Repairing/replacing broken slabs
   - Reshaping and compacting canal embankment
   - Sealing leakages and illegal diversions
   - Cleaning and reshaping laterals
   - Clearing and reshaping drains

b. Dam wall, spillway and reservoir
   - Checking and refilling eroded portions
   - Planting grasses at the down stream slope
   - Replacing rip-rap (boulders) on the upstream slope
   - Clearing dam wall of regenerated trees and shrubs
   - Refill eroded portion of the spillway
   - Removing dangerous aquatic weeds

c. Catchment area protection
   - Enforcing no farming regulation within the catchment area
   - Prevent bushfires within the catchment area
   - Planting and maintaining trees and grasses
   - Reshaping contour bunds where they have been constructed

Farmers mobilize cash and labor resources and carry out the maintenance of the irrigation schemes, but when the expertise needed for a particular activity is

---

[24] Svendsen and Huppert (2003) define irrigation system maintenance as a technical activity and a service aimed at keeping irrigation infrastructure at a desired performance capacity or to restore it to a particular capacity.

beyond the capabilities of the WUA, technical experts are hired. Cash for maintenance is raised from irrigation fees as well as levies for specific activities (Section 4.5). Irrigation fees, however, remain the main source of cash income for the associations surveyed. Although the stated willingness to make additional cash contributions towards maintenance was high among the WUAs, little evidence exists to show that such contributions have been made in the past. The level of resource mobilization demonstrates the willingness of the beneficiary groups to contribute to collective action for the operation and maintenance of the schemes. In addition to the irrigation fees members contribute a substantial amount of labor for maintenance activities. On the average, members contribute 23 hours of work (2.9 work days) per year in labor for collective maintenance activities. This figure compares with 3.3 workdays for a group of irrigators in Mexico reported in Dayton-Johnson (1999).

Maintenance work is carried out collectively with all members working side by side. Formal announcement of impending work is made to all members. Households in principle contribute equal amounts of labor and cash for maintenance, irrespective of the size of their plot (or benefits). If a member has a genuine reason for being unable to participate, he is obliged to provide an alternative person to himself; either a household member of 15 years and above or hired labor.

It was observed that maintenance of the irrigation system (network of canals and drains) was the sole responsibility of the irrigators. However, the entire WUA and in some cases the community as a whole come together to undertake maintenance of the dam infrastructure and the protection of the catchment area due to the interest some communities have in the sustainability of the dam for its multiple uses (section 4.7). In Figure 4.3 we present the distribution of community labor for maintenance activities at the community irrigation schemes.

Figure 4.3: Distribution of community time for maintenance

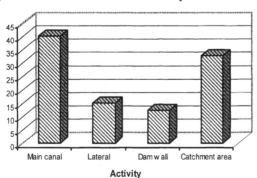

4.4.1 Maintenance of canals and other structures

Many of the WUAs have developed maintenance manuals that require them to undertake regular walkthrough inspections to ascertain the health of the irrigation systems and to inform themselves of the interventions required. Some of the system maintenance activities are routine, some periodic and others performed under emergency often requiring technical advice. Emergency works require immediate action to prevent or reduce the effect of an unexpected event such as breach of the dam embankment. The routine maintenance activities include cleaning of canals, laterals and drains of weeds and silts. Periodic maintenance carried out to get systems ready for cropping include sealing of leaks in lined canals, replacement of broken slabs, and reshaping and compacting of canal embankment.

Maintenance activities carried out on the dam infrastructure (e.g., dam wall) are largely preventive measures. Incentives for carrying out preventive maintenance can, however, be weakened especially where rights over the physical facility are ambiguous. Maintenance of canals, laterals and drains are activities frequently undertaken by the WUA, with approximately 55% of total maintenance time spent on these activities as depicted in Figure 4.3 above.

4.4.2 Catchment area protection

Siltation of the dam resulting from human activities (including farming) within the protected catchment area had been a major cause of the breakdown of the irrigation infrastructure in the study area in the past. Consequently, catchment area protection has become an important component of recent programs designed to promote the sustainability of the schemes. The amount of maintenance time (33%) spent on catchment area protection (as shown in Figure 4.3 above) reflects the general understating of the need to protect the catchment area. Indeed the WUAs were expected to enact bylaws that would seek to protect the catchment area from encroachment and perennial bushfires, and to mobilize labor for the planting of trees and grasses to check erosion of the area and silting up of the reservoir. However, catchment area protection remains one of the problematic areas threatening the success of the community-based organizations in the management of the schemes. In some cases, farming is taking place within the designated catchment area while the nurseries of tree seedlings established do not extent beyond what the project staff established for demonstration. The survival rates of seedlings are very low due to the difficulty of protecting the young trees from grazing livestock, especially during the dry season when the irrigation areas appear to be the only sites with green vegetation. Planting of vertivar grasses is being pursued but the success rates vary across schemes.

There is no gainsaying the fact that the sustainability of the schemes greatly depends on the ability of the communities to protect the catchment area, particularly to prevent bushfires and farming in the area in order to reduce erosion and siltation of the reservoirs. Construction of contour bunds in the catchment area, as envisaged by the LACOSREP, will go a long way to help in this direction. A key constraint to catchment area protection, however, lies in the capacity of the WUA to enforce its bylaws in connection with the catchment area. Given that most of the catchment area has long been considered as "government land", many people in the villages are not motivated to stop encroachment in the area. More education and sensitization may help to change this perception and to save the schemes.

Maintenance of the irrigation system, as outlined above, is a labor intensive activity that requires greater cooperation from members. The ability of the WUA to undertake the seasonal repairs and improvements is usually limited by labor, money and other resources it can mobilize during the season. Beyond this, the communities usually have little choice but to depend on the central government or NGOs, and may remain trapped in a cycle of insufficient maintenance and declining performance. In the following section we examine the performance of the user groups in mobilizing funds (irrigation fee collection) to finance operations and maintenance.

4.5   Levies and funds mobilization

The water users' associations raise funds through water levies to carry out operation, administration and maintenance activities that involve monetary outlay. In general membership dues and water levies are decided by the WUA during a general meeting. Initial membership fee (3000-5000) is often charged, after which most of the WUAs levy annual dues (water charges). In the newer WUAs, the initial membership contribution is deemed to be a share capital deposited in the WUA fund to cover part of expenses towards cooperative registration and future maintenance cost (Roy – pers. Comm.).

Due to the lack of water gauging devices irrigation levies are not based on the amount of water one uses but often on flat charges per person per season. There were attempts by few WUAs to charge fees based on plot size (number of vegetable beds) but that resulted in some problems due non-uniformity of bed sizes, in addition to other measurement problems. Flat rates charged range from ¢2000 in some WUAs to ¢30000 per person per season in other schemes. It also came to light that in most of the communities (78%) there were no separate fees for livestock owners and fishermen neither do they pay levies. Often levies and modes of payment are instituted by the user groups based on local conditions. Three levy payment schedules were observed; full payment before harvesting (in 36.5% of the groups), full payment after harvesting (48.1%) and partial payments (15.4%) where some percentage of the levy is accepted before gardening begins and the remaining amount paid after harvest. These schedules reflect liquidity constraints that farmers face during some periods of the year and past experience of the WUA with regards to fee collection.

Figure 4.4: Levy collection performance

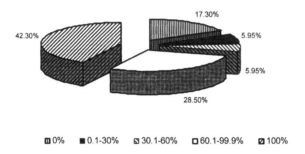

⊞ 0%   ■ 0.1-30%   ☒ 30.1-60%   ☐ 60.1-99.9%   ☒ 100%

Fee collection performance (measured by the percent of members who have paid their dues in full, as at the time of the survey, i.e., end of 2002/2003 dry season) was generally above average. The mean percentage fee collection performance was 70.8, ranging from 0 to 100%. In particular, 42.3 percent of the WUAs achieved 100% collection while 17.3% recorded zero as depicted in Figure 4.4 above. This indicator can, however, be affected by the economic conditions of the irrigators and the degree to which the irrigators feel the system is worth supporting. Nevertheless, lack of transparency and accountability in the use of WUA funds has affected the ability of some WUAs to improve on their levy collection performance. Field discussion with some WUAs revealed that sometimes it takes the threat of sanctions to impel some of the defaulting members to pay up.

4.6   The survey households

4.6.1 Characteristics of the sample households

In this study, the unit of analysis of individual activities at the schemes is the household. The sample was therefore carefully selected to avoid interviewing more than one member of the irrigators association from the same household. Coincidentally, many of the respondents turned out to be the heads of their respective households. Table 4.4 presents selected demographic characteristics of sampled households. About 80% of our respondents were males, 88% of which were heads of their respective households. The majority of the female respondents were heads of their households, many of whom were widows, and it is likely that the married ones have been made de facto heads because their husbands have migrated for off-season job opportunities in southern Ghana. Household sizes are large, ranging from as low as 1 to the largest of 34 persons.

Table 4.4: Characteristics of sampled households in northern Ghana

Characteristics of households	Frequency	Percent
Female		
Age distribution of household heads		
Less than or equal to 25 years	69	13.3
26- 35 years	128	24.6
36 – 45 years	131	25.2
46 – 55 years	86	16.5
56 – 65 years	62	11.9
More than 65 years	44	8.5
Education of head (years of completed)		
0	333	64.0
1-5	46	8.8
6-10	101	19.4
10+	40	7.7
Household size		
1 – 4	38	7.3
5 – 9	278	53.5
10 – 14	131	25.2
15 – 19	47	9.0
20+	26	5.0
Primary occupation of head		
Agriculture	497	95.6
Trading	5	1.0
Fishing	1	0.2
Handicraft	4	0.7
Wage Employment	11	2.1
Other	2	0.4
Number of observation = 520		

A typical household is composed of a man, his wives and children and grandchildren. The parents of the man, his brothers and other extended family members may be residing with him either temporarily or permanently. The mean household size was about 9.6% compared to the national average (4.9) and 5.55 for rural savannah (GSS, 2000). Mean household dependency ratio in the sampled households was 0.4 (ranging from 0 to 0.83).[25] The external family system in the study region makes single person households (0.19%) negligible. Small household sizes of 1-4 persons constitute 7% of the sample while households with sizes ranging from 5-9 persons represent about 54%. Larger

---

[25] A household's dependency ratio is calculated by dividing the number of individuals under 15 years of age plus the number of individuals over 65 years of age by the total number of individuals in the household. A dependency ratio of zero implies a household whose resources are not strained at all, while the ratio of one indicates a household whose resources are extremely strained.

households (10 or more persons) make up about 36% of the WUA members interviewed.

Table 4.4 also shows the age distribution of the heads of the households, which ranged from as young as 18 to as old as 95 years. The largest share of members of the irrigators association in our survey is found to be in the age group 25-55. This identifies with earlier studies (e.g., Cutler, 1999) that tend to suggest that middle aged groups are more identified with groups that require contributions from members. The exact peak age however varies from study to study, but ranging from 35 to 55 years (see Smith, 1994). The mean age of the sampled household heads was 43 years.

Education provides a more flexible atmosphere for collective management. It allows sufficient room for new ideas and enhances the ability to profit from different information sources. However, a large proportion of our sampled household heads (64.2%) cannot read nor write. Largely, the illiterate population has limited opportunities to earn alternative employment outside agriculture. It is therefore expected that the majority joins the WUAs to enable them to earn some income through dry season gardening. Only 35.8% of the household heads have ever attended school, out of which about 64% attained six or more years of formal education. About 19% of the household heads completed 6-10 years of education. The mean of years of schooling was 2.9. Significantly, the level of education attained by the household heads declined with age.[26] See also Table D4.1.

4.6.2 Household income sources and asset distribution

Agriculture is the main source of livelihood for the majority (96%) of the survey households. The households undertake both crop cultivation and livestock rearing. Despite being the source of livelihood, agriculture hardly sustains many households. Seasonal food shortage is persistent. Often, the proceeds from the main (rainy) season agriculture are not enough to guarantee a household's livelihood. As a coping strategy, many of the respondents undertake off-season economic activities to supplement their incomes. Young men migrate out of their communities at the end of the farming season in October in search of economic opportunities in southern Ghana, and return at the beginning of the farming season to cultivate their fields. Although migration can have a negative effect on resource mobilization for maintenance activities, it has un-intended impact of increasing female WUA membership, and possibly long-term usufruct rights of land for women (IFAD, 2002). A large number of the respondents (69%) reported secondary occupation, while 25% undertake tertiary economic activities to earn additional incomes for the household. For men the main secondary economic activities include trading in livestock, handicraft (basketry, weaving, etc.), and fishing while very few are salary earners. Women on the

---

[26] The correlation coefficient of age versus years of education = -0.265 (0.000, 2-tailed).

other hand engage in handicraft (including pottery), pito brewing, shea-butter processing, fuel wood gathering, and petty trading. More women (69%) engage in handicraft than men. Indeed, the majority of the households in the sample (72.9%) engage in off-farm economic activities.

In spite of the myriad of income generating activities the respondents undertake, income from dry season gardening constitutes a substantial share of annual cash income for many of the respondent households. Over 60% of the respondents reported that more that a half of their household's income is obtained from irrigation activities undertaken in the dry season.

The main staple crops (sorghum, millet, maize, groundnut, and others) are produced by farmers under rainfed conditions in the wet season, mainly for home consumption. The dry season production on the other hand is largely meant for the market (75%), with the greater percentage of the proceeds going into the purchase of staple crops to supplement household production in the wet season. Though some irrigation communities are highly noted for the cultivation of particular crops, our survey revealed that a varying amount of the major vegetables is cultivated across the communities in all the surveyed districts (section 2.3.1).

Table 4.5 presents some information on asset ownership and distribution. Our survey gathered information on all types of assets owned by the households including: houses, land, livestock, farm equipment, as well as other productive and nonproductive assets. Productive (agricultural) assets comprised of agricultural land, bullock, plough, and other items used on the farm, all valued in Ghanaian cedis (¢). Buildings, vehicles, furniture, bicycles, motor-bikes, radios, sewing machines, and other household items constitute the non-productive (non-agricultural) assets. The average value of all assets was about ¢14.0 Million. This value excludes agricultural land because there was no consistent information on land prices as land in most of the rural communities is never sold or leased out (section 2.6.3). This phenomenon is, however, not the case in the urban communities where land markets exist (especially for residential purposes). The mean values of assets owned by male and female headed households were respectively ¢16 Million and ¢8.8 Million. Productive assets constitute about 60% of all assets owned by sampled households. There was no statistically significant difference in means of non-productive assets held by male and female headed households.

Livestock held by households in our sample included cattle (including bullock), donkeys, sheep, goats, pigs and poultry. As predicted, female headed households owned fewer livestock (mostly small ruminants) than males. Property rights in livestock can be complex with implications for ownership and distribution patterns in some parts of the study area. Indeed, it is common in some of the ethnic communities that women and young men within the households cannot have cattle in their name. All cattle in the household are held by the head who also takes all decisions about the use to which the livestock is put. However, the extent to which this restricted property right affects our sample was not investigated. Almost every household in our sample owns poultry. Many of the households derived some income from the sale of livestock. During the hunger period (April-July), which also coincides with the peak growing season and high demand for labor, households sell livestock often at very low prices to purchase food and to pay for farm inputs. Small ruminants (especially, sheep and goats), pigs or poultry are often sold to meet immediate cash needs of the households.

Table 4.5: Asset ownership and distribution

Asset Characteristics	Sample	Male	Female
*Mean value of assets (¢)^*			
All assets	14,000,000	16,000,000	8,823,832**
Productive farm assets[a]	8,513,892	9,473,957	4,523,178***
Non-productive assets[b]	5,599,669	5,914,226	4,300,663
Livestock	5,558,949	6,232,924	2,762,953***
Mean Agricultural land owned (ha)			
Upland (non irrigation land)	3.3	3.6	1.8**
Irrigation plot	0.14	0.16	0.09***
% Households owning	**%**	**%**	**%**
No land	1.3	1.9	4.0
Less than 1 ha	19.0	16.5	29.7
1 – 5 ha	71.9	73.7	64.4
5.01 – 10 ha	4.2	5.0	1.0
10.01- 15 ha	1.7	1.4	0
Greater than 15 ha	1.3	1.4	1.0
% Size of irrigation plot			
Less than 0.05ha	24.6	21.5	37.6
0.05-0.099ha	7.3	8.1	4.0
0.1-0.49ha	56.4	58.2	48.5
0.5ha or more	11.7	12.2	9.9
No. of observations	520	419	101

a. farm equipments, bullocks and other items of used for the farm, but excluding land;
b. buildings, furniture and other household items
^Ghanaian Cedi (¢). At the time of survey ¢8600 = 1US$
The difference in means between male and female headed households was tested; *, **, ***
represent level of significance at 10%, 5% and 1% respectively.

Ownership and control of land follow a similar pattern as in the distribution of households' physical assets. The mean size of land (upland) was 3.3ha, but land sizes are smaller in the female-headed households (Table 4.5 above). Field discussions revealed that females in the study area have no right to land, but widows may have temporal control over land which they hold in trusts for their growing male children. The land distribution trend shows that about 90% of households own less than 5ha of land. Very few households own more than 10 hectares of land.

The policy to accommodate as many community members interested in dry season gardening as possible has resulted in the fragmentation of the irrigation land into tiny plots. The sizes of plots cultivated by the respondents during the 2002/2003 dry season range from 0.008 – 1ha, with more women than men cultivating less than 0.05ha. The mean plot sizes were 0.16 and 0.09 for male and female headed households respectively. The difference between the means is statistically significantly different from zero, supporting the observation that females hold smaller plots than males. The mean plot size, for the sample, is 0.14 (Coefficient of variation, C.V. = 1.13). The majority of the respondents (56%) cultivated between 0.1 – 0.49ha. The high coefficient of variation in plot sizes suggests some degree of inequality in plot distribution across the schemes. The general perception among respondents was that original landowners own the larger plots while many of the gardeners with smaller plots were women.   The tendency for powerful members (landlords, executive members, etc.) to grab more land to themselves undermines the equity objective of the promotion of community-based resource management.

The pattern of inequality in plot allocation is depicted in figure 4.5 below.[27] The left hand side panel depicts plot inequality across the survey districts in the Upper East and Upper West regions. Plot sizes are relatively less unequal in Wa and Nadowli districts of the Upper West region and more unequal in the Builsa, Bwaku West and Bawku East Districts in the Upper East region. The average Gini coefficient for plot distribution is 0.35; the maximum Gini is 0.72, the minimum is 0.03.[28] However, the extent of inequality varies across communities. In the right hand side panel, plot inequality at the community level is depicted (see Appendix A for list of communities). Plot distribution is particularly unequal at Googo, Bugri and Paga Nania.

Figure 4.5: Inequality in irrigation plot allocation

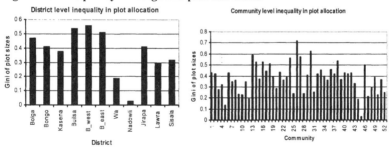

---

[27] Gini coefficients of inequality in plot allocation were calculated for each community and for each district, using DAD 4.5 for distributive analysis (Danclos, Araar and Fortin, 2002).

[28] The overall Gini coefficient for the sample is 0.6.

Inequality can have significant influence on the level of cooperation for collective maintenance of the schemes. However, theoretical predictions about the impact of inequality on collective action remain ambiguous (Section 5.3).

4.6.3 Social organizations

Inter-household network of relationships has increasingly become important in studying collective action for the management of natural resources. The level of social interaction and mutual dependence among members affect the ability of communities to act collectively. Cooperative interaction on any aspects of social or economic activities of community members leads to development of norms, networks and trust because of which people act for common welfare. Households belonging to many local organizations get better sensitized about the benefits of collective action and are more likely to cooperate in collective maintenance activities.

The majority of the sampled households, however, do not belong to any other local associations other than the water users' association. Among those associating with other social organizations, 82% belong to only one association while the remaining 18% belong to two or more groups (Table 4.6). Only one person belonged to four social organizations. It is expected that WUAs in which most of the members belong to other social organizations which undertake regular civic activities are more likely to succeed in the governance of the schemes. Among the prominent informal groups to which many of the sampled households are associated are savings and credit association, farmers' associations (other than WUA), women's groups, religion-based groups, and others including youth association and hometown-based/ethnicity-based groups. Close to 13% of the household heads were members of farmers' associations (Table 4.7). Nine percent (also 48% of the female respondents) are members of women's groups. Some 29% of the respondents have served as members of community development committees of their respective towns and villages.

Table 4.6: Number of informal groups in the survey communities

Number of groups	Frequency	Percent
0	313	60.2
1	170	32.6
2	30	5.8
3	6	1.2
4	1	0.2

Table 4.7: Types of informal groups in the survey communities

Informal groups	Frequency*	Percent
Saving and credit groups	21	4.0
Farmers' associations	67	12.9
Women's groups	48	9.2
Religious based associations	17	3.3
Other	20	5.0

*Multiple response

Many of the women's groups undertake joint off-farm economic activities such as malting, brewing, shea butter extraction, and handicraft. This feature enhances group dynamics and promotes accumulation of knowledge and experience in self-organizations which can be useful for a durable community-based organization for the sustainable management of the common-pool resources.

4.7 Multiple uses of irrigation water

Although the dams were primarily constructed to support dry season gardening, the water is used for many other purposes. Understanding the multiple uses of the schemes is essential for the development of management strategies that meet the needs of the multiple users (Yoder, 1981, 1984). The main uses of water from the reservoirs include consumptive uses (such as livestock watering, gardening, drinking and home construction), non-consumptive uses (such as fishing, washing/laundry and bathing) as well as environmental and ecological functions (e.g., supply of biodiversity in pants and wildlife, especially migratory birds). In particular, the dam reservoirs are the main sources of water for livestock in the study areas during the dry season, when most of the water sources (seasonal streams) dry up. Irrigation, however, constitutes the largest consumer of the dam water. It is estimated that about 55% of the reservoir volume is available for irrigation, while only 8-10% is used for domestic and livestock consumption. While other consumptive uses may take relatively little water, they may have a high value in the estimation of the communities, which implies that focusing on only the irrigation field crop may lead to underestimation of the value of the irrigation systems (Meinzen-Dick and Backer, 2001).

As illustrated in section 3.5.2, the management structure of the WUA identifies the major stakeholders involved in the multiple uses of the irrigation water as the core management groups. It defines the types of access and use rights each group has and these use rights would be affected by changes in the amount of water in the reservoir. In particular, the WUA, as the platform for negotiating water uses, is made up of irrigators, livestock owners, and fishermen. It must be recognized that the present management structure is meant to provide a forum that would deal with the problem of multiple uses.

In many communities, the dam water increases per capita water availability throughout the year, as it remains a major source of domestic water supply available to many households in the study area. The importance of non-agricultural uses of irrigation water in livelihood strategies has implications for irrigation management and water rights as increasing water scarcity challenges existing water allocation mechanisms. In Table 4.8 we present the uses to which the dam water is put among the surveyed communities.

Table 4.8: Uses of the dam water in the surveyed communities

Main Uses	Frequency*	Percent
Gardening	51	98.1
Fishing	41	78.8
Livestock	52	100
Domestic	18	34.6

* multiple responses

Vegetable gardening, the most important irrigation water user, is identified as the main dry activity in most of the communities surveyed, and the source of income for the majority of the people as outlined in the preceding sections. The establishment of the irrigation schemes is therefore seen as an important intervention to reduce out-migration of the youth and associated social problems.

Livestock is the most important enterprise in the area. Livestock watering is the single most important use of the dam water that extends the benefits of the scheme to all members of the beneficiary community, as every household tries to keep livestock (section 2.3.1). Livestock kept include cattle, donkeys and small ruminants. Indeed, northern Ghana has the largest livestock concentration in Ghana and these are all traditional herds that are grazed freely. The important roles livestock play in the livelihood strategies of households make water a critical need not only for household use but also livestock watering. The difficulty of finding water for the livestock during the dry season was a concern. It is in this respect that the intervention of government and non-governmental organizations in the construction of dams and dugouts in the study area has been very crucial. The concern of the community in getting water for the livestock to reduce the risk of livestock loss through theft, which was a major occurrence in the past when the animals stray far away in search of water, motivates interest in the maintenance of the infrastructure.

Water use rights for livestock are clearly defined, but cattle watering places were not recognized in the scheme design. However the systems' design took into account the need to always reserve some amount of water for livestock. In many of the dams the dead storage of the reservoirs was raised during rehabilitation to keep the water level in the reservoir at a certain minimum to ensure that a substantial amount of water is always left for livestock, fisheries and domestic use.

Although gardening is the largest consumptive use of the dam water, priority of use changes in times of water scarcity. In many communities pre-season negotiations take place between stakeholders, as to whether or not to crop in the season, judging from the level of water in the reservoir. In years of poor precipitation and very low water level in the reservoir, use rights shift in favor of livestock and fisheries. This can have implications for management especially in communities where almost every household (members and non-members of the WUA) keeps some livestock.

The reservoirs also serve as fish habitats. In some of the schemes deliberate efforts were made by the fisheries department of the Ministry of Food and Agriculture to promote aqua-culture. The rehabilitated dams were stocked with fingerlings and the beneficiary communities assisted with technical expertise to manage the fish stock. Interested members of the communities as well as fishermen constituted themselves into associations to regulate fishing. Indeed, fishing rights are restricted to the fishermen associations which also pay levies to damsite management committees (DMCs).

One other important function of the dam water is in the area of domestic uses. Domestic uses of the dam water include cooking, bathing/laundry and brick-making for home construction. Though water taken from the dams is not treated before consumption many of the communities depend on the dams as water sources especially in the dry season when many water sources dry up. Water use planning by the association does not take account of domestic uses. Perhaps this attitude is influenced by the fact that the dead storage volume is also expected to take care of domestic requirements. No representation exists on the DMCs for domestic water uses and no fees are fixed for domestic water uses.

However, the multiple use functions of the dams in most of the communities encourage the traditional leaders and development committees of the communities to show much interest in the sustainability of the schemes. To most of the communities, the dam is a common property that must be protected, an interest that influences communal participation in the maintenance of the dam infrastructure. An example of this sort of communal interest was evident at Winkongo where the chief had organized the community to collect clay materials to be used to manage seepage in the dam wall. At Dua, also in the UER, the chief's interest in ensuring that enough water is always available for the livestock and domestic uses drives him to discourage gardening in bad rainfall years when the reservoir level is low. Thus, the multiple uses of the irrigation facility serve as the greatest incentive for the members of the community to participate in collective maintenance of the schemes. However, as a community property, it is open to the problems of collective action (e.g., cooperation and free-riding).

Clearly, multiple uses can be a source of motivation for sustainable management, but these can also be sources of conflicts, which could hinder cooperation for maintenance of the facilities if the institutional arrangements (including organizations and norms) do not take into account the interests of all stakeholders.

In the remaining chapters, we attempt to find out how the characteristics of the user group, resource attributes and household characteristic, as well as institutional factors affect the success of collective action for the management of the community irrigation schemes.

5   THEORY AND CONCEPTUAL FRAMEWORK

5.1   Introduction

Many aspects of community management of common-pool resources (e.g., irrigation systems) require collective action on the part of user groups in the provision and maintenance of common goods such as water infrastructure, catchments area protection, etc. Collective action can be reflected in the willingness of groups of people to contribute to the provision of common goods (e.g., the construction of collective infrastructure). It can take the form of people's participation in the setting up of regulatory agencies endowed with powers to collect fees, impose contributions on members, lay down rules and sanction violators. Collective management requires cooperation. Cooperation not only requires action by the community to define rules but also commitment to monitor and enforce these rules. Considering that these actions involve costs, the interested parties will cooperate whenever the benefits of cooperation exceed the costs (Ostrom, 1992b; Wade, 1987).

Following Olson's (1965) exposition on the free-rider problems different authors have examined the issues in different ways. Socio-anthropological case studies and game theoretic models have been used to explain cooperation among agents in natural resource appropriation.[29] Case studies basically offer a contingent explanation for the phenomenon of collective action, but they do not permit testing for the relative importance of hypothesized factors, and their generalizability to other cases is unclear. Many scholars have also used game-theoretic models to debate the outcomes of collective action. Game theoretic models tend to provide analytical explanations to the phenomena, but they are often associated with a multiplicity of equilibria, making it difficult to resolve many comparative static questions satisfactorily without recourse to contextual data analysis (Baland and Platteau, 1996). Besides, these models are often set up in a framework that is too restrictive and inadequate for capturing some of the important issues influencing real world cooperation (Bardhan, 2000). Though game-theoretic models have virtue of precision, it is only empirical evidence that can determine potential relevance of any particular formulation (Dietz et al., 2002). This study makes a contribution in this direction by collecting data on a sufficiently large number of irrigation communities in northern Ghana and conducting an econometric analysis of the factors hypothesized in the literature on collective action and local management in an attempt to explain differences

---

[29] Examples of case studies include Ostrom (1990), Bardhan (1993), Wade (1988), Lam (1998), Ostrom and Gardner (1993), Ostrom, Gardner and Walker (1994), Runge (1986). Examples of game theoretic models include Kreps et al. (1982), Axelrod (1984), Sugden (1986), Hirshleifer and Rasmusen (1989), Weissing and Ostrom (1991), and Baland and Platteau (1997, 1998, 1999). For a review of this literature see, for example, Baland and Platteau (1996), and Rasmussen and Meinzen-Dick (1995).

in resource management outcomes across communities. The main advantage of this approach is that it permits an empirical assessment of the validity and relative importance of hypotheses derived from socio-anthropological case studies and game theoretic models for the specific case of irrigation management in northern Ghana. We draw on the existing literature to develop a conceptual framework for the analysis and to derive hypotheses on the factors determining performance differences across communities.

5.2 A Survey of selected literature on commons management

A large number of both theoretical and empirical literature on commons management have emerged since the publication of Hardin's (1968) article on the "Tragedy of the Commons". The literature related to this important subject is so extensive that the following pages could not cover all the available material. Agrawal (2001), Baland and Platteau, (1996), and Hardin (1992) present comprehensive review of some of the literature on the problems of commons management.

In the literature, the concept of tragedy of commons has been used to explain overexploitation and degradation of natural resources (Stevenson, 1991). Hardin and others argued that common property regimes were inefficient and could result in overexploitation, degradation and eventual ruin of collectively used resources as a result of the user's rational incentive to maximize his own benefits (Stein and Edwards, 1998). Indeed, Hardin's conclusion that "freedom of the commons brings ruin to all", advocating for private or government ownership of the commons, motivated policy changes especially in developing countries in the 1960s and 1970s that transferred natural resources from their previous property rights regimes to government control (see Arnold and Campbell, 1986). The prevailing belief of this school of thought was that the problems associated with common-pool natural resources could only be solved either through privatization of the resource or through state intervention (Smith, 1981; Johnson, 1972; Hardin, 1968).

In many publications about common-pool resource management, however, Hardin's ideas are criticized (see for instance, Ostrom and Walker, 1994; Ostrom, 1990; Wade, 1987). According to the critics, Hardin unfortunately used the term "commons" to refer to "open access", although they differ essentially in terms of decision-making arrangements that are present to govern their use. The confusion perhaps stems from lack of clarity between open access and common property resource use regimes. Ciriacy-Wantrup and Bishop (1975) note that common property is not everyone's property. The concept presumes a well defined user group and that potential resource users who are not members of the user group are excluded. Essentially, open access refers to a 'free for all' situation, where ownership of the resource is not defined, and so there are no rules to control access and use of the resource units. Common property regimes, on the other hand, are characterized by a structured ownership within which a set of rules governing access and use of the resources are developed and enforced, and incentives exist for co-owners to follow accepted norms (Ostrom, 1990; Wade, 1987; Oakerson, 1986). Thus, the presence of an institutional structure is very important, without which common property would not be differentiated from open access. Table 5.1 summarizes private, common property and open access resource use regimes based on group and extraction limitations.

Table 5.1: Types of resource management regimes

	Private Property	Common Property	Open Access
Group limitation	One person	Members only	Open to all
Extraction limitation	Extraction limited by individual decision	Extraction limited by rule	Extraction unlimited

Source: Stevenson, 1991

In the light of this, Hardin's "Tragedy of the Commons" often results from lack of institutions and failure of social mechanisms such as communication and trust, and ability to make binding agreements and enforce decisions (Stevenson, 1991; Bromley, 1991). Essentially, members involved in joint exploitation of common property resources face the dilemma of whether to compete for a bigger share from the resource or cooperate to increase the resource pie through collective action[30] (Van der Linden, 1999). It is postulated that collective management of common-pool resources by users could be an appropriate system for ensuring sustainability and for that matter overcoming the tragedy of the commons (Ostrom, 1992, 1990; Berkes, 1989; Wade, 1987). Rappaport (1984) argues that the tragedy of the commons could be averted by mechanisms that can cause individuals to act in collective interest.

In a common-pool resource scenario, collective action will typically occur if local stakeholders seek to overcome the problems associated with (as Stevenson, 1991 puts it) 'the tragedy of open access', and agree on decision-making arrangements to control access and use of common-pool resources. But the logic of individual rational utility seekers may not coincide with the logic of the community (Olson, 1965).

Olson (1965) shows that the organization of groups to pursue collective ends, was vulnerable to a paradox often referred to as the "free rider"[31] problem, such that actions by individual members may worsen rather than enhance collective wellbeing. In Olson's (1965:2) opinion "rational self-interested

---

[30] Collective action is described as action by a group (either directly or on its behalf through an organization) in pursuit of members' perceived interests (Oxford Dictionary of Sociology).
[31] A free-rider is a social actor who perceives that he will receive a higher individual payoff for a socially defecting choice than for a co-operative choice, even though all individuals engaged in collective action would, in the long term, be better off by working together.

individuals will not act to achieve common or group interest". The scholar asserts that all group interests were subject to the same social dilemma, connected to tragedy of the commons and the problem of prisoner's dilemma, because players are tempted to free ride on the cooperation of others given that there is always a benefit from defecting. That is, when benefits cannot be withheld from non-contributors, rational individuals will not voluntarily contribute to public goods, but will be motivated to free ride on the contributions of others. It was argued that due to the temptation for individuals to free ride for selfish gains collective action can only be achieved through some selective incentives.

Kimber (1981), however, argues that Olson's 'rational analysis" is itself illogical and internally inconsistent. For it rests on the assumption that only the free-rider is rational, and all others willing to create the public good that the free rider can consume without cost are irrational. He argues that free-riding and tragedy of the commons certainly exists, but that these behavioral patterns are not necessarily universal. There can be similarly rational basis for more altruistic and cooperative behavior (Upholf and Langholz, 1998; White and Runge, 1994). Indeed, empirical evidence has emerged over the past decades that suggests that local user groups are often capable of managing common pool resources through collective action through clearly defined property rights (Bromley., 1992; McCay & Acheson, 1990; Ostrom, 1990; Wade, 1988).

Ostrom (1990) notes that an appropriately designed property rights system can help user groups to overcome the collective action problem. She shows that in many situations people are able to cooperate to improve joint outcomes. Ostrom asserts that, the problem facing user groups of the commons is that of organizing in order to supply and maintain institutions[32] that will promote cooperative behavior. Collective action is thus possible under circumstances that involve adherence to agreements specifying actions of each individual. The institutional arrangements established to promote cooperative behavior is thus very crucial for the success of collective action. It is argued that Hardin's tragedy of the common often results not only because of the failure of the common property arrangement but also from the failure of institutions to control access to and use of resources (Berkes and Folke, 1998; Dove, 1993).

In the context of common-pool resources, Richards (1997) defines institutions as a set of accepted social norms and rules for making decisions about resource use. These define who controls the resource, how conflicts are resolved, and how the resource is managed and exploited.

---

[32] Institutions can be defined as formal and informal rules and norms which govern or at least influence the behavior of participants of a society in their social, political and economic interactions, (and commonly understood as what actions are required, permitted, or forbidden in particular situations.(Shaffer, 1969) Thus, Institutions, being collective conventions and rules provide a surety for the continuous flow of interactions and therefore guarantee the stability of any property rights system (Manig, 1991).

Ironically, different authors have different views about how institutions emerge. Literature which emphasizes design principles (see for instance Ostrom, 1990; Wade, 1988; Baland and Platteau, 1996, etc) connotes functionalist and normative approach to institutional development. Underlying such models is the importance of productive and distributional concerns in determining the incentives to cooperate with institutional arrangements (Cleaver, 1998). The functionalist thoughts on common pool resource management are gradually being replaced by donor assisted interventions rooted in norm-based controls (Campbell, 2001).

It is the view of the evolutionist that institutions for natural resource management can be crafted well only by resource users but not policy makers (Ostrom 1992a). Rejecting the assumption that external actors can efficiently craft durable institutional solutions and enforce rules, Ostrom (1992a) argues that users of the commons are better equipped to resolve the cooperation problems. Rules and constraints are perceptions commonly known to participants and used to organize repetitive interdependent relationships. Ostrom views 'crafting' as a continuous evolutionary process of developing the optimal institution to support collective action that produces a public good and to discourage abuse. Culture and social structure then become the raw materials or resource bank from which institutions can be built to promote cooperation in resource management (Cleaver, 1998). In line with the induced innovation hypothesis (Kikuchi and Hayami, 1980; Hayami and Ruttan, 1985; Binswanger and McItyre, 1987), Hayami and Kikuchi (1980) argue that collective regulation of a resource may evolve when the resource becomes scarcer and when its privatization is prohibitively costly. Hayami and Kikuchi point out that under such conditions "the social structure becomes tighter and more cohesive in response to a greater need to coordinate and control the use of resources as they become increasingly more scarce" (Hayami and Kikuchi, 1980: 21-22).

Cleaver, however, points out the fact that the evolution of collective decision making institutions may not be the process of conscious selection of mechanisms for the collective action task (as Ostrom assumes in her model) but rather the outcome of individuals acting within the bounds of circumstantial constraint. Thus, the incentives to cooperate are based on local organizational capacity, appropriate cost-benefit sharing arrangements, empowerment of resource users and a complex and diffuse reciprocity occurring over lifetimes (Cleaver, ibid). He questions the possibility of consciously and rationally crafting institutions for collective action and supports instead ideas about multiple processes of institutional formation combining both conscious and unconscious acts, unintended consequences, and a large amount of borrowing of accepted patterns of interaction from sanctioned social relationships.

So far, the predominant models of institutions in common property resource management literature are essentially bureaucratic, emphasizing on clearly structured arrangements for decision making often involving representation, regularization and formalization (Nelson, 1995; Ostrom, 1990;

Cleaver, 1998b). Often times, traditional, and informal institutions are considered inherently weak and there is a common assumption that modern arrangements can make good the deficiencies in the traditional mechanisms of achieving cooperation (Seabright, 1993, in Cleaver, 1998). However, the growing opinion is that the presence of a well-established set of decision-making arrangements is not enough to guarantee collective action in the long term. The existence of institutions that can be adapted for new purposes may be extremely important for the sustainability of self-governance and successful management of common-pool resources (Ostrom, 1990).

Oakerson (1992) operationalizes institutional arrangements as three sets of variables: Operational, structural and performance variables. The operational variables comprise of rules which affect the use of the resource directly. These include allocation rules, monitoring and sanctioning rules, incentive structure, etc. Structural variables on the other hand refer to the nature of the collective choice rules, including the structure of the user association, decision making procedures, etc. while performance variables covers private benefits and livelihood improvements. The extent to which these arrangements affect the performance outcomes of local management common-pool resources will be explored in this study. First we review the factors the factors that have been hypothesized in the literature to condition successful collective action and local management of natural resources.

5.3   Hypothesized conditions for successful commons management

Much of the bourgeoning literature on commons management that have emerged, over the years, examine the problems of collective action and focuses on getting understanding of what motivates people to participate in collective action for the management of the commons (see for instance, Pinkerton and Weinstein, 1995; Ostrom, et al 1994; Tang, 1992; Sengupta, 1992; see also Hardin, 1982; Baland and Platteau, 1996, and Agrawal, 2001, for reviews). Ostrom (1990) presents eight design-principles which seem to be favorable to sustainable use of common-pool resources. Wade (1988) finds 14 conditions to be important in facilitating successful management of the commons. Baland and Platteau (1996) also derive seven factors underlying successful collective action from empirical evidence, although theirs overlap to some extent with those of Wade and Ostrom. Table 5.2 highlights the factors identified by scholars of the commons as significant in facilitating the sustainable management of common pool resources.

These factors have often been discussed under four categories: physical characteristics of the resource, characteristics of the user group, institutional arrangements, and external factors. The relative importance of these factors as drivers of collective action is, however, determined by local conditions. In this review we summarize the predominant conclusions made about these factors in the literature. See Agrawal (2001) for a comprehensive and synthesizing review of the studies by Wade (1988), Ostrom (1990) and Baland and Platteau (1996).

Table 5.2: Factor facilitating successful local management of the commons

1. Resource system characteristics
(i) Size of the resource
(ii) Clearly-defined boundaries
(iii) Levels of mobility
(iv) Possibilities of storage of benefits from resource
(v) Predictability

2. User group/Community characteristics
(i) Group size
(ii) Clearly-defined boundaries
(iii) Prevalence of shared norms
(iv) Presence of past successful experiences/Social capital
(v) Leadership/Local hierarchies
(vi) Heterogeneity in endowments
(vii) Heterogeneity in identities and interests
(viii) Interdependence among group members

1 and 2. Relationship between resource system characteristics and group characteristics
Overlap between user group residential location and resource location
Levels of dependence by group members on resource system
Fairness in allocation of benefits from common resources
Nature of changes in levels of user demand

3. Institutional arrangements
(i) Locally vs. externally devised access and management rules
(ii) Degree to which rules are simple and easy to understand
(iii) Ease in enforcement and monitoring of rules
(iv) Availability of low cost adjudication
(v) Accountability of monitors and other officials to users

4. External environment
(i) Cost of exclusion technology
(ii) Time for adaptation of new technologies related to commons
(iv) Levels of articulation with external markets
(v) Nature of changes in articulation with external markets
(vi) Central government undermining of local authority
(vii) External sanctioning institutions
(viii) Levels of external aid to compensate local users for conservation activities

Source: Extracted from Agrawal (2001)

Generally, it is hypothesized that characteristics of the resource such as size, boundedness (clearly defined boundaries and users), resource scarcity, dependence on resource, etc., affect community resource management outcomes. The size of the resource, for example, has largely been hypothesized to have a positive relationship with collective action (see for example, Ostrom, 1990; Wade, 1988). It is however suggested that the relationship between resource scarcity and collective action resembles an inverted U, as little cooperation is expected during abundance or extreme scarcity (Bardhan, 1993, Tang, 1992). Focusing on irrigation schemes, Bardhan and Tang argue that both too much and too little of (for example) water can hamper collective management of irrigation schemes. On the one hand, under conditions of extreme scarcity, institutional arrangements for cooperation are less likely to be successful (Bardhan, 1993). On the other hand, high rainfall or availability of private wells may compel individuals to break out of collective action arrangements (Tang, 1992).

Similarly, Gibson (2001) shows that people will not attempt to preserve resources that they do not depend on or perceive as valuable. It is argued that a high level of dependence on resources in a subsistence-oriented environment is likely to be associated with better governance of common pool resources (Wade, 1988; Varughese and Ostrom, 2001). However, other studies suggest that dependence on the resource may have an inverted U-shaped relationship with cooperation in local commons management (see Agrawal 2001a). When the incomes individuals derive from the common resource decline to below subsistence levels while alternative sources are not available, the temptation to violate access and use rules are likely to be irresistible due to survival constraints (Baland and Plateau, 1996).

The discussion on characteristics of the user group emphasizes the socio-economic conditions of the community as determinants of resource management outcomes. These factors include the size of the user group, heterogeneity, prior experience in community mobilization, social capital, and appropriate leadership.

The effect of the number of resource users on collective action has been much debated. An earlier work by Olson (1965) suggests that collective action is easier to organize in smaller groups as peer monitoring is easier in small groups, while shared norms and patterns of reciprocity compel users to consider the indirect and long-term consequences of their actions (see also Poteete and Ostrom, 2003; Baland and Platteau, 1999; Bardhan, 1993; Ostrom, 1990; Bromley, 1992; Wade, 1988). It is postulated that effective rules are less likely to be sustainable when group size is large.

Olson's group size argument has been criticized (see Hardin, 1982, for a review of this debate). Critics argue that larger group size does not have to lead to underprovision of the collective good.

Baland and Platteau (2000) illustrate the impact of group size on collective provision the collective good. They show that large groups that share common norms and are faced by common challenges may be successful. Besides, there may be some positive economies of scale in large groups in the matters of pooling resources and sharing risk.[33] The ambiguity in Olson's argument about group size has been commented on by other scholars who assert

---

[33] Also, Chamberlin (1974) and Isaac et al. (1993) observe that if the cost of monitoring and sanctioning remains the same, then increasing the group size provides additional hands for sustaining the resource for the benefit of all.

that the relationship between group size and collective action is not very straight forward, but mediated by other factors such as production technology, heterogeneity, size of resource, etc., (Agrawal and Goyal, 2001; Ostrom, 1997; Marwell and Oliver, 1993). The relationship between group size and collective action is an area that requires further research.

One other factor that has received considerable attention in the analysis of collective action is group heterogeneity. Baland and Platteau (1996) emphasize the importance of interdependence among group members based on mutual understanding for building institutions that would promote sustainable resource management. Heterogeneity among agents in a society reflects the extent to which trust influences the emergence of local management, and affects incentives for lasting cooperation (Ostrom, 1999; North, 1990). Zak and Knack (2001) posit that heterogeneous societies with weak formal and informal institutions have lower trust and retarded economic performance than less heterogeneous, higher trust societies. Communities may be heterogeneous in economic (endowments, interests, etc.) or socio-cultural background (or identity), and these may affect cooperation in collective action differently (Baland and Platteau, 1999, 1996). It is argued that heterogeneity based on identity (social, ethnic, political, gender, etc.) can create obstacles to cooperative behavior (Bardhan, 1999; Baland and Platteau, 1999) because it can hamper communication and impede the development of trust and cooperation among group members. Sharing social, cultural and economic characteristics may increase the predictability of interactions which can in turn provide the basis for trust to facilitate collective action. Moreover, heterogeneity can limit the opportunities available to weaker members to participate in and benefit from decision making processes (political economy considerations, see Engel, 2003), as participation in community decision making can be skewed in favor of powerful individuals (Hissan *et al*, 1991).

The effect of economic heterogeneity (e.g., inequality in wealth, inequality in land holding, inequality in exit options or other endowments) on cooperation is, however, less clear. Empirical evidence on this aspect remains ambiguous (Baland and Platteau, 1999; Kambur, 1992). One school of thought holds that inequality based on endowments, (e.g., wealth or land distribution) is good for collective action, as privileged members with greater interests can take initiatives for collective action to provide the common good even if the poorly endowed group chooses to free ride (Agrawal 2001b; Hardin, 1992; Olson, 1965). Thus, it is suggested that heterogeneity emanating from differential endowment of users may possibly enhance cooperation when it does not hinder uniformity of interest in collective agreement and the privileged can assume leadership roles and provide authority structure needed for the enforcement of regulatory rules (Agrawal, 2001a).

Other scholars, on the other hand, argue that economic and social heterogeneity hamper collective action (see for instance, Johnson and Libecap, 1982; Jayaraman, 1981, Easter and Palanisami, 1987). In situations of marked inequality some individuals may not contribute to the collective action and the result may be inefficient outcomes. Enforcement cost of cooperation may also be high in ethnically heterogeneous and conflict prone situations, because the extent of confidence and trust among resource users may be low. Ostrom *et al* (1984) suggest that when individuals do not trust one another, cannot communicate efficiently, and cannot develop agreements, then outcomes are more likely to march theoretical predictions of non-cooperative behavior among rational individuals playing finitely repeated complete information games (see also Agrawal, 2001a). Bergstrom, et al., 1986 predict a neutral effect of economic inequality on collective action.

Vanghese and Ostrom (2001) link the effect of heterogeneity on collective action (either positive or negative) to the specific form of organization and institutions established for the management of the resources. The results of studies (e.g., Dayton-Johnson and Bardhan, 2002) depicting a U-shaped relationship between inequality and collective action add to the growing ambiguity in the existing literature. Discerning empirical studies that link heterogeneity to better or worse outcomes of local management is still lacking. In particular, the extent to which certain types of heterogeneity (social, economic, etc) enhance or undermine collective action remains unclear. Our study attempts to make a contribution to these issues as well.

The external environment, including social, institutional, technological and demographic factors affect collective action although limited attention has been accorded to it in literature. External factors like market integration, access to credit, and population growth are also hypothesized to affect collective action in resource management (Baland and Platteau, 1996; Bardhan, 1993; Myers, 1991; Ostrom, 1990).

The impact of market integration on collective action has been widely debated (Agrawal, 2001b). On the one hand, market integration is postulated to have a positive relationship with cooperative management as sites closer to markets are likely to be more commercially oriented, and for that matter have higher payoffs to effective cooperation in irrigation management (e.g., Agrawal and Yadana, 1997). Other views hold that market integration erodes the ability of local organizations to successfully manage the commons (Baland and Platteau, 1996; Young, 1994). Fujita, Hayami and Kikuchi (1999) point out that in rural societies with little exposure to urban market activities, community members expect to continue their interactions forever, and hence have incentives to cooperate. Access to markets often decreases this interdependence and creates different incentives and exit opportunities capable of reducing the need for cooperation in collective management. Also market access may decrease the incentive of members to abide by community rules, by increasing the opportunity cost of labor or providing more exit options, and for that matter rendering rule enforcement weaker (Pender and Scherr, 1999; Balland and Platteau, 1996). Moreso, increased access to markets raise the opportunity cost of labor and opens new and more rewarding opportunities in the case of irrigation breakdown, making it less optimal for users to contribute time and resources towards collective management of the commons.

Variations in the levels of local population depending on the common pool resource, can significantly affect the ability of users to create rules to manage resources. Hayami and Kikuchi (1981) are of the view that growing resource scarcity due to population growth can motivate the emergence of collective organizations that will promote conservationist practices. However, Dasgupta (1995) argue that high population growth rate negatively affects the likelihood of collective action, as this can make enforcement of rules very difficult. Community rules are likely to dissolve under intense pressure of population growth. People are inclined to violate sharing rules to appropriate large shares, when the individual's share of the resource become small and livelihood cannot be supported from it (Baland and Platteau, 1996). Besides, conservation strategies that also include population reduction may not work as it can weaken local institutions (including collective labor arrangements). Emigration can also reduce labor availability for maintenance (Garcia-Barrios and Garcia Barrios, 1990).

The state also has an important role to play in the functioning of common pool resource management arrangements by providing communities and resource users legal backing required for their rules and regulations to be enforceable (Rangan, 1997). Gibson (1999), Ribot (1999), and Richards (1997), among others, examine the complexities of state-local relationships, yet we do not have clear understanding of the relationship and how this affects outcomes of common-pool resource management. Besides, gradual usurpation and erosion of the authority and functions of the traditional leaders by the state make them loose legitimacy, while the spread of western education and values lead the new generation to question old forms of traditional authority and power, often resulting in reduction in potency of traditional measures of social control. Devolving authority (with legal backing) to traditional institutions can thus help to resolve equity concerns and promote efficient management of resources at the local level.

The review of the existing literature shows that there is still a great deal of debate over the factors that determine the success of collective action. The direction of the effect of many variables on collective action remains ambiguous in the general literature. It is also worth noting from the foregoing review that most of the design principles, including those examined in this section, are expressed more as definite measures required for successful commons management rather than as factors that depend for their efficacy on the presence or absence of other variables. Understanding local institutional arrangements and their interactions with the resource system will help to make sense of the ambiguous associations reported in the literature.

Broadly, the existing literature suffers from some important and closely related weaknesses. Firstly, though there is extensive knowledge about the determinants of success of common-pool resources management systematic empirical evidence on the subject for large samples remains limited. Only few studies vigorously test the relative importance of factors identified in the literature as conditioning the success of local management of common-pool resources through large-scale sample econometric analysis (see, Dayton-Johnson, 2000; Heltberg, 2001; Meinzen-Dick, *et al*, 2002).

Secondly, the existing literature has tended to focus more on key factors such as resource system and local groups. Very little attention is paid to social and political dynamics that shape interactions in local communities. This prevents the emergence of better understanding of how the list of the hypothesized factors interacts with local institutional arrangements and resource systems (Cleaver, 1998). Third, existing studies on local commons have largely emphasized on the outcomes of community-based management in terms of sustainability of resource use. Other types of outcomes, such as the economic efficiency of local resource management have not been explicitly analyzed. Also not very clear is the condition under which local management of resources can ensure equitable distribution of costs and benefits among resource users.

Moreover, the factors listed in Table 4.2 are often treated in the literature as exogenously determined variables in the analysis of factors influencing collective management of common-pool resources, ignoring endogenous relationships that in reality exist between most of these variables (Engel, 2003; Baland and Platteau, 2002; Agrawal, 2001; Heltberg, 2001). As Baland and Platteau (2002: 65) observe "it is indeed difficult to deny that some important characteristics of the resource user group are not given parameters but variables over which the users themselves have some degree of control". In reality, many of the factors are endogenously determined in the process of devolution and community-based resource management. In particular, users may want to define their own profile and organization to make them more conducive to a collective mode of regulation (Engel, 2003). The importance of defining causal links between the various factors has also been stressed by Agrawal (2001a). Yet many of the existing studies have not been availed to this methodological challenge. In the following section, we present a framework that conceptualizes the endogenous effects in the analysis of collective action.

5.4   Conceptual framework of analysis

Although a large number of potentially critical factors have been identified as conducive for local management and sustainability a complete theory that explains sustainable common-pool resource management is yet to be developed (Agrawal, 2001a). In Figure 5.1 we present an analytical framework for the study, based on the literature on collective action and country specific factors based on expect interviews. It considers five groups of variables: the physical attributes of the resource system, characteristics of the user group, household incentives, external factors, and collective decision-making and outcomes, and incorporates the fact that some of the variables discussed in the literature are endogenous, i.e., they are determined within the community decision processes. The hypothesized signs (directions of effect) of these factors on the outcomes of community-based resource management are as discussed in the preceding section. The common presumption in the literature is that external factors such as public policy, population pressures, access to market, etc., affect decision-making at all levels; the resource system, community level characteristics and household incentives. The attributes of the resource itself together with community level factors shape the opportunity costs and constraints that irrigators face. Together with household incentives these influence collective decisions shaped by appropriate institutional arrangements, which in turn determine outcomes. On the other hand, collective decisions and outcomes can be determining factors in explaining community and household incentives.

The hypothesized relationship between the set of variables and how we intend to conceptualize our empirical analysis are presented in figure 5.1 below.

Figure 5.1: Conceptual framework

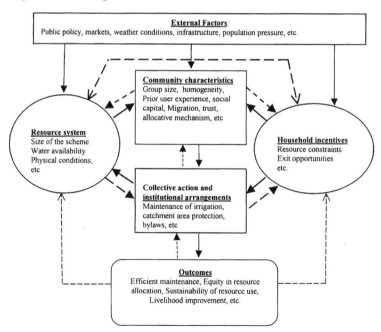

In the context of this study, equity and sustainability outcomes of collective action are assessed by analyzing the following:

- Rules for distribution of costs and benefits
  - Equity in the distribution of irrigated land
  - Equity in water distribution
  - Conflicts over land/water
- Rule conformance
  - Institutional supply
  - Levy payment
  - Water poaching
- Scheme maintenance
  - Maintenance of canals
  - Dam wall protection
  - Catchment area protection
  - Quality of maintenance

As Figure 5.1 illustrates, many interaction effects and endogeneities exist between the separate outcomes (e.g., profitability of irrigation and land distribution are important outcomes in themselves, but they are also important factors hypothesized to determine irrigation maintenance activities). Below, a theoretical model, based on Sandler (1992) and Baland and Platteau (2002), is presented to portray the endogeneity effects in the analysis of collective action.

5.5   Conceptual model

Variants of the public good model have been used in the analysis of cooperation on common property resources, including irrigation schemes. In the following section, such a model is presented to demonstrate how individual farmers' decisions on their contribution to resource management may lead to free riding and a suboptimal provision of the common good (e.g. irrigation maintenance). Cooperation for collective action can potentially overcome this problem. It is postulated that the success of managing the resource collectively by the community will be a function of benefits and cost of providing the management. Thus, community members will pool their efforts together to follow corporate management whenever the net benefits to be derived to all or most members of the group are high. The model also provides a conceptual basis for our earlier statement that some amount of dependencies exists among the variables, which can result in endogeneity problems (Engel, 2003; Baland and Platteau, 2002; Agrawal, 2001; Heltberg, 2001).

The basic framework for analyzing collective action problems have been set up by Olson (1965), Buchanan (1968), Sandler (1992), and Cones and Sandler (1996). Our analysis follows that of Sandler (1992), Baland and Platteau

(2002) and Rueben (2003) to assume a user group in a community whose members cooperate to manage an irrigation system. We assume further that the system consists of N farming households (agents) $i = 1, 2, ..., n$ who make positive contributions to the maintenance of the collective good (in our case irrigation infrastructure). Specifically, individual $i \in N$ contributes $g_i \geq 0$. Total level of maintenance, M, is given by $M = \sum_{i=1}^{n} g_i$. The payoff of participation in maintenance is assumed to be a function of benefits associated with maintenance quality and cost of participation. i.e., individual i gains utility $U_i(M)$ from the use of the irrigation scheme (collective good). $U(\cdot)$ is assumed to be twice differentiable such that

$$U_i(M) > 0 , \ U_i''(M) < 0 , \ \forall \, M \geq 0. \tag{5.1}$$

The utility gained by the entire group $U(M)$ is the sum of the individual utilities $U_i(M)$. The cost or utility loss of each individual for producing the collective good is given by the cost function $C_i(g_i)$. We assume that $C(\cdot)$ is twice differentiable with

$$C_i(g_i) > 0 \ \text{and} \ C_i''(g_i) > 0, \ \forall g_i \geq 0. \tag{5.2}$$

The Nash equilibrium in such a collective action requires that farmer i chooses his/her contribution (effort) g_i to the provision of the common good to maximize utility. Thus, we formulate the utility maximization problem by assuming that the individual contributes a positive amount to the collective good only if the utility he/she gains exceeds the utility loss (cost);

$$U_i(M) > C_i(g_i) . \tag{5.3}$$

Thus, each individual in the group faces the following maximization problem;

$$\underset{g_i}{Max} \, U_i(M) - C_i(g_i) \tag{5.4}$$

$$\text{s.t. } U_i(M) > C_i(g_i)$$

In a Nash behavior each individual treats the optimizing contribution level of the rest of the group as given when maximizing his/her objective function. That is, household i in taking its decision, takes the decision of all other households as exogenously given. That means that the equilibrium levels of all households are in principle related to each other, but g_j is considered constant by i when taking its decision.

Thus, each member of the group maximizes his utility by solving

$$\underset{g_i}{Max} \, U_i(g_i + \sum_{i \neq j} g_j^*) - C_i(g_i) \tag{5.5}$$

Thus, under this assumption the best response g_i^* of individual $i \in N$ (assuming both the first order condition and the second order condition hold) is given by:

$$\frac{\partial U_i (g_i^* + \sum_{i \neq j} g_j^*)}{\partial g_i} = \frac{\partial C_i (g_i^*)}{\partial g_i} \tag{5.6}$$

i.e., when all the individuals use the best responses, the solution to the problem for every $i \in N$ is

$$\frac{\partial U_i (M^*)}{\partial g_i} = \frac{\partial C_i (g_i^*)}{\partial g_i} \tag{5.7}$$

Thus, the individual will produce the collective good to the point where the private marginal benefit equals the marginal cost. It also holds that $g_i^* = 0$ if

$$U_i (M^*) \leq C_i (g_i^*). \tag{5.8}$$

The simultaneous evaluations of first order conditions (5.7) for all households yields the Nash equilibrium level g^* for each i, the sum of which gives the Nash quantity M*, the provision level that results in the absence of cooperation or collective action.

Sub-optimality results at the Nash equilibrium because the individual equates his marginal benefit to the marginal cost ignoring the benefit that his/her contribution bestows on others. I.e., he/she does not take into account the effect of his contribution on the utility of other group members when deciding contributions. To show this, consider the maximization problem the group will face if it acted together. In this case any Pareto optimal contribution level must maximize the aggregate net benefit and therefore must involve the level of contribution that solves

$$\max_{g_1, \dots, g_n} U(M) - \sum_{i=1}^{n} C_i (g_i) \tag{5.9}$$

$$\text{s.t.} \quad U(M) > \sum_{i=1}^{n} C_i (g_i)$$

The necessary and sufficient first order condition for the optimal level of contribution by each is

$$\sum_{j=1}^{n} \frac{\partial U_j (M^{\circ})}{\partial g_i} = \frac{\partial C_i (g_i^{\circ})}{\partial g_i} \tag{5.10}$$

where g_i° is the community optimal level of g_i that maximizes $U(G)$. Thus, equation (5.10) shows that the group will cooperate when its marginal benefit equals marginal cost. This implies that if the user group cooperates and chooses collective action to maximize common utility, a Pareto optimizing quantity M° can be obtained. However, since

$$\frac{\partial U_i(M)}{\partial g_i} < \sum_{j=1}^{n} \frac{\partial U_j(M)}{\partial g_i} ,$$ (5.11)

$\forall M > 0, i \in N$ and $C_i''(\cdot) > 0$, $\forall i \in N$,

it follows that, g_i° must be strictly greater than g_i^*, and for that matter M° is strictly greater than M^*.[34] As depicted in the Figure 5.2 below, the optimal level of the public good chosen by the community as a whole (given by solving equation (5.10)) exceeds the level of the public good resulting from individual utility maximization,[35] thus affirming Olson's (1965) under provision hypothesis.

Figure 5.2: Suboptimality of individual provision

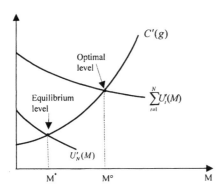

Source: adapted from Mas-Colell et al. (1995).

Intuitively, in the absence of collective action, individual households only consider the benefits of their provision to the common good on their own utility, but not the benefit on the utility of other households nor the spillins of the public good from the contributions of others. Thus, the lack of cooperation leads to an underprovision of the common good. To overcome this problem user groups

---

[34] Exception is when costs are so high that even for the group it makes no sense to produce the collective good (see also Rueben, 2003).

[35] The curve corresponding to the $\sum_{i=1}^{n} U_i'(M)$ geometrically correspond to a vertical summation of the individual curves representing $U_i(M)$ $\forall i$.

need to find a way to cooperate, and devise, monitor and enforce effective management rules.

However, collective action will be difficult to achieve if the required effort to achieve improvement far outweighs the expected benefits. Furthermore, collective action will be less successful in achieving desired outcomes where there are inequities in the distribution of costs and benefits and institutions for ensuring compliance to rules are weak. Baland and Plateau (2002) argue that expected benefits of collected action (defined here as the level of maintenance) is dependent on the attributes of the resource system (e.g., size of the resource, boundedness, water supply conditions, etc,) and other gains associated with resource appropriation (e.g., social capital, benefits from catchment area protection, etc.). Similarly, the cost elements capture maintenance costs and the user group's ability to efficiently manage the system by defining rules, monitoring behavior of members and enforcing rules, etc (McCarthy, Sadoulet and de Janvry, 2001). Consequently, the cost of collective action is said to be a function of the characteristics of the user group (e.g., group size, homogeneity, social capital, exit opportunity, etc.), the attributes of the resource, characteristics of the group, as well as external factors including public policy relating to the devolution process.

Incorporating the above arguments into the utility function and solving (5.10) while taking into account the issue of endogeneity associated with some of the variables will yield the following system of simultaneous equations:

$$Mt = f(Dist, RC, \Pi, AR, CC, GC, Inst, Ext) \qquad (5.12)$$

$$Dist = f(Mt, RC, \Pi, AR, CC, GC, Inst, Ext) \qquad (5.13)$$

$$RC = f(Mt, Dist, \Pi, AR, CC, GC, Inst, Ext) \qquad (5.14)$$

$$\Pi = f(Mt, Dist, RC, AR, CC, GC, Inst, Ext) \qquad (5.15)$$

where: Mt = Quality of Maintenance; Dist = rules for distributing costs and benefits; RC = Rule conformance; Π = Profitability of irrigation; AR = Vector of attributes of the resource system; CC = vector of community characteristics; GC = vector of group characteristics; Inst = vector of institutional variables; and Ext = vector of external factors.

Conformance with community rules (e.g., regarding the payment of water levies and the provision of labor for maintenance work), profitability of irrigated agriculture and distribution of cost and benefits clearly affects the resulting irrigation maintenance activities at the community level. Equity in the distribution of cost and benefits affects cooperation in maintenance and rule conformance which also affects profitability. On the other hand, the aggregate maintenance activities also affect profitability of irrigated agriculture. Similarly, conformance with rules regarding the payment of water levies or the provision of labor for maintenance affects households' costs and thereby the average profitability of irrigated agriculture. Finally, rule conformance is the result of each household's decision to comply with community rules or not, and thus depends on the perceived benefits and costs, which in turn depend on the profitability of irrigation and the level of maintenance.

Clearly, the multiple endogeneities and interactions between variables in the model developed would ideally require estimating system of simultaneous equations. However, the size of our data presents some limitations in the estimation procedures that can be used. In chapter 6 we specify empirical models for the analyses of data while dealing with the endogenous relations described above.

Households decide to contribute maintenance effort, the aggregate amount of which affects community maintenance performance. Therefore, in addition to the above structural relationships used to analyze the community level outcomes, we also utilize a household labor allocation decision model to examine factors that condition household allocation of labor for the maintenance of the community irrigation schemes. The outcomes of collective action at the community irrigation schemes are analyzed in chapter six.

Finally, the empirical analysis of irrigation management outcomes in Northern Ghana requires that the main hypotheses from the literature (summarized in section 5.3) are adapted to local conditions and potentially some country- or region-specific hypotheses are added.

6 COLLECTIVE ACTION IN THE FRAMEWORK OF COMMUNITY IRRIGATION MANAGEMENT IN NORTHERN GHANA

6.1 Introduction

Devolution of the management of natural resources from government agencies to resource users has often been justified based on the conviction that users have a comparative advantage and self interest over government agents in managing the resources. However, for the policy to succeed in achieving sustainability of the irrigation systems, it will require collective action among the user group members for operation and maintenance. Though the theory of common pool resources has been advanced significantly, it remains uncertain how resource users will be able to overcome the cooperation dilemma and supply themselves with successful institutions. Understanding the causes of success or failure of the collective action strategies can provide valuable lessons for strengthening efforts to promote sustainable user management of irrigation schemes and other common-pool resources.

The previous chapters have highlighted the structure of the water users association and collective activities undertaken by the user group to manage the irrigation systems. In this chapter, we analyze the collective action issues in an attempt to explain why some communities are achieving better outcomes than others in the management of the schemes. In particular, section 6.2 analyzes the collective action issues at the community level. In the framework of a household labor allocation model the incentives for households to participate in the collective maintenance of the irrigation facilities are analyzed in section 6.3.

6.2 Collective action for irrigation management.

In this section, we analyze three forms of collective action visible in the survey schemes: the provision of a collective good, distribution of costs and benefits, and concerted action by a user group to comply with rules regulating the use of the common-pool resource. In particular, we examine how the following forms of collective action are undertaken by the communities and the determinants of community effectiveness: (i) maintenance of the irrigation schemes, (ii) rules for distribution of costs and benefits, and (iii) conformance with rules (i.e., compliance with payments of irrigation levy and compliance with water allocation rules).

Maintenance tasks undertaken by the Water Users Associations (WUAs) have been examined in section 4.4. Consequently, maintenance issues addressed in this chapter relate to the quality of maintenance produced by the user groups through collective action. But first, we discuss the rules for the allocation of costs and benefits and subsequently the equity implications of the distributive regimes on the success of collective action (measured by the quality of maintenance achieved through internal cooperation).

6.2.1 Rules for distribution of costs and benefits

The construction of dams to promote dry season gardening has been an important aspect of poverty alleviation programs being implemented by the government of Ghana and NGOs in the study area. The poverty alleviation objective can, however, be achieved only when the majority of the members of the beneficiary communities, especially the poor, have fair opportunities to benefit from the irrigation schemes. In this sub-section we examine the allocative rules that apply in the schemes studied and the associated equity implications. As stated earlier, maintenance costs - in terms of both labor and cash contributions – are shared equally and are allocated on a per-household basis. Thus, the allocative rules of interest here particularly define how the available resource (irrigation land and water) is to be distributed among the members of the user groups.

In section 6.2.1.2, we examine the reasons why distributive conflicts emerge in some communities, but not others. As the choice of water distribution rule is thought to be one of the potential explanatory factors, we will, however, first address this issue. The land allocation function of the WUA has been discussed in section 3.6.

6.2.1.1   Water distribution

Individual irrigators, by virtue of their membership of a WUA, have a right to equal share of the irrigation water for gardening, making the task of water distribution an important task in the operation and maintenance roles of the WUAs (Yoder, 1994). Often the distribution policies are based on the need to promote equity in water rights to users. The conversion of the water rights to day-to-day water distribution is, however, a sensitive issue that can result in inefficient and unfair distribution, a potential source of distributive conflict (Dinar et al., 1997).

Across the survey schemes, the water distribution mechanisms were decided by the WUA through mutual understanding. Field observations showed that water distribution arrangements in many instances deviate from the distributive mechanisms prescribed to the WUAs. The users have altered the prescribed distribution arrangements to meet their expectations. The WUAs appoint among its members someone to keep the valve keys and oversee the opening and distribution of water in accordance with pre-arranged schedules. In most of the schemes, however, this task is carried out by chairpersons of the associations.

Variations of irrigation water distributive rules have been identified in the literature. These include time rotation, depth of water, area of land, and share of flow (See Yoder 1994). The WUA carries out this responsibility without outside interference. However, the rules for distributing water among the irrigators are similar across survey schemes but vary between time rotation and continuous flow arrangements. Under the continuous flow system of water distribution water is opened to all canals and channels at the same time, for each farmer to draw water to his/her field. In some of the schemes where continuous flow arrangement is practiced, water opening was irregular and there was no planned schedule. Often the amount of water supplied to individual plots depends on the location of one's plot and the length of time the flow is allowed to last. So that those with plots located at the tail end of the scheme tend to receive less water than those close the dam wall (head enders).

On the other hand, under the rotational arrangement water is taken in turns according to a predetermined schedule specifying days, times and duration of supply to each section in the command area. Turns rotate from one section of the scheme to another. Because the schemes are generally smaller in sizes, complete cycles last a day or two. Issues such as the beginning and ending of irrigation or user sequencing in the rotation mechanism are decided by the WUA through mutual understanding.

The need to use scarce water judiciously and economically has been the concern of the irrigation development authority. The rotational system of water distribution has been promoted as an efficient and equitable means of managing irrigation water, especially when rainfalls in the area are low and erratic and water levels in the reservoirs often low, despite doubts about the validity of the system as an efficient method of water allocation to meet crop water requirement (Reidinger, 1980). However, the system can result in disagreements if a means is not found to prevent water poaching that arises out of the tendency for other users to divert water onto their fields when it is not their turn. In many communities where the rotational system is used farmers spend much time tracking water poachers and blocking channels that have been opened illegally. Indeed, the ability of user groups to conserve water depends on the local norms and the strength of institutions to enforce these norms. If the user group does not promote efficient use of water, distribution rules alone will have little effect on demand management (Dinar et al, 1999).

None of the systems studied has water distribution structures, the absence of which makes it difficult to directly address equity issues relating to water allocation. However, water distribution arrangements adopted by the associations, to use Ostrom's (1992) expression, are clearly crafted to suit their conditions. Generally, water distribution is based on the principle of equal division, but in principle it is in proportion to shares of landholding in the irrigation area. The amount of water each farmer receives depends on the size of his/her plots, with bigger-plot owners getting proportionally larger amounts of water. This suggests incongruence between cost sharing and water allocation rules, a situation which has some implications for the success and sustainability of self-management regimes. It has been argued in the literature that a mixed rule not only discourages maintenance (Dayton-Johnson, 2000a; Olson, 1992), but may also reduce land conflicts (Engel, 2003).

Baland and Platteau (1999) have observed that institutions governing the use of resources have distributional implications that can be critical to their success. Incongruence between rules for sharing costs and benefits can have implications for collective action. Maintenance burdens requiring larger investments (effort and money) than returns to farmers may not lead to sustainable farmer management. Demand for equity on the part of the aggrieved farmers can lead to conflicts. Thus, water distribution regimes adopted can have a significant influence on cooperation for collective action. Distributive regimes can however be endogenous, affected by structural characteristics. In section 6.2.1.1.1, we attempt to analyze the choice of water allocation arrangements.

6.2.1.1.1 Choice of water allocation arrangement

Dayton-Johnson (2000b) uses theory and econometric analysis to explore how self-governed Mexican irrigation societies choose distribution rules, and concludes that the choice of water allocation rule is greatly influenced by the age of the water users association and inequality in land holdings. Using our dataset from northern Ghana, we attempt to test this and other underlying hypotheses. Recalling from the conceptual model presented in section 5.5, rules for distribution of costs and benefits are assumed to be a function of the attribute of the resource system (AR), characteristics of the community (CC), user group characteristics (GC), institutional factors (Inst), External factors (Ext), as well as some endogenous variables such maintenance (Mt) and rule conformance (RC)[36]. Specifically,

$$Distributive\ Rule = f(Mt,\ RC.,\ \Pi,\ AR,\ CC,\ GC,\ Inst,\ Ext.) \qquad (6.1)$$

which reflects the choice the irrigating community has to make between the two water distribution arrangements. Variables are as defined under equations (5.12) - (5.15). To analyze the decision making processes of the user group, a discrete choice regression model is employed. This enables the estimation of the community decision as a binary dependent variable. The binary choice here relates to the need for the community to make one of two choices; continuous flow or rotational mechanism of water distribution. Thus, the probability of the

---

[36] In an attempt to resolve the problems of endogenous relationships in the model, 2-stage procedures (where variables are generated in first stage regressions) are employed in this and subsequent analyses presented in this Chapter.

user group in community i selecting a distributive regime (say continuous flow) is formulated as a binary choice D* that is linear in exogenous variables. In a simplified manner;

$$D_i^* = x_i'\gamma + u_i; \quad u_i \sim N(0,1) \tag{6.2}$$

where D_i represents the distributive regime; x is a vector of explanatory variables; γ is a vector of coefficients to be estimated; and u_i is the error term. D_i^* is a latent representation which is not observed. Thus, instead of observing D* we observe a binary variable indicating the sign of D_i^*:

$$D_i = \begin{cases} 1 & if \ D_i^* = x_i'\gamma + u_i > 0 \\ 0 & if \ D_i^* = x_i'\gamma + u_i \leq 0 \end{cases} \tag{6.3}$$

For our example, we observe the continuous flow arrangement whenever the statement is D=1, and 0 otherwise (the rotational arrangement is observed). Thus, the probability of interest is

$$P(D_i = 1) = P(D_i^* > 0) = P(x_i'\gamma + u_i > 0)$$
$$= P(u_i > -x_i'\gamma) = 1 - \Phi(-x_i'\gamma) = \Phi(x_i'\gamma) \tag{6.4}$$

The likelihood function used to estimate the probability can be stated as:

$$L = \prod_{y=0}[1 - \Phi(-x_i'\gamma)]\prod_{y=1}[\Phi(-x_i'\gamma)] \tag{6.5}$$

Maximizing the log likelihood function with respect to γ gives the estimates of γ. Depending on the assumptions made about the error term, the most common models used in the analysis of binary response variables, logit model and probit model, can be derived. Whilst the logit model assumes a logistic distribution, the probit model assumes a normal distribution of the error. However, the two distributions are more or less similar except that the logistic distribution has a fatter tail. The estimated results of the probit and logit models also do not differ much rendering the decision to select between these two models a matter of convenience. We have selected the probit model for further analysis, thereby assuming that the error term (u_i) follows a normal distribution. The probability of observing a continuous flow distribution regime in community i, is written as

$$P(continuous._i) = Prob(D_i = 1) = \Phi(-x_i'\gamma) = \int_{-\infty}^{x_i\gamma} \frac{1}{(\Pi)^{1/2}} e^{\frac{-t^2}{2}} dt \tag{6.6}$$

where t is a standardized normal variable with the mean zero and variance equal to one. Thus, the probit model of the continuous flow arrangement is specified as:

$$P(D = 1) = x_i^{'}\delta + \varepsilon_i \qquad\qquad (6.7)$$

6.2.1.1.2 Variables and hypotheses

The factors used to explain the choice of the water distribution arrangement included the number of villages using the scheme, water supply conditions, age of the user group, ethnic heterogeneity, social organizations, quality of maintenance, profitability, size of user group (WUA size), water conflicts, size of the scheme, "illegal" extension of the scheme by the community, and resistance of landlords to land redistribution. However, quality of maintenance, rule conformance and water conflicts can be endogenous to the system. Proxies and variables created in first-stage regressions are used in place of the endogenous variables. The predicted values for profitability and quality of maintenance (Tables D6.1 and D6.2) are used in this analysis. Existence of village-wide rules against non-participation in communal work is used as a proxy for rule conformance in user groups, whilst prevalence of chieftaincy disputes in the area served as a proxy for water conflicts.

    The WUAs of schemes shared by many villages as well as those in larger groups may choose the rotational arrangement of water distribution to promote equity in water distribution. However, it may be difficult for a large number of villages and a larger group of users to agree on a water allocation arrangement that involves a large degree of coordination and monitoring. Therefore, the signs of the variables 'number of villages' and 'groups size' are ex ante ambiguous. For a similar reason the sign of scheme size is ambiguous. Under conditions of frequent water shortages, WUAs are expected to choose the rotational system of water arrangement. The signs of ethnic heterogeneity and conflicts are also a priori ambiguous as it may be difficult in heterogeneous and conflict-prone societies for the WUAs to agree on distribution arrangements. On the other hand, the user groups are expected to employ rotational systems to ensure equity and promote harmony in conflict-prone areas. It is expected that older WUAs choose the rotational system of water allocation due to past experiences with water shortages. In communities where landlords resisted land redistribution, the WUAs might prefer an allocation system that would enable them to assert their authority control over the system, i.e., give them greater control and legitimate authority over the system. Continuous flow arrangement will be expected in communities where rules are frequently violated. Communities with experiences with social organizations are more likely to be socially cohesive and more able to develop better monitoring systems to reduce abuse of water allocation rules. Such communities are expected to choose a rotational regime for distributing irrigation water.

    Table 6.1, however, presents the definitions and descriptive statistics of the variables used in the analyses presented in this chapter.

Table 6.1 Description of variables included in regression equations

Variable	Variable definition	Mean	Std
Scheme characteristics.			
Scheme size	Size of irrigation area in ha	12.16	7.93
Water shortage	1= if experienced frequent water shortages; 0 otherwise	0.54	0.50
Quality of rehab.	1= if community was satisfied with quality of rehabilitation work; 0 otherwise	0.29	0.45
Labor intensive	1= if scheme rehabilitation by labor intensive techniques; 0 otherwise	0.44	0.50
Dam extension	1= if dam has been illegally extended by the community; 0 otherwise	0.59	0.49
Maintenance*	Predicted values of quality of maintenance (log)	-1.21	0.44
Group characteristics			
WUA size	Membership strength of water users (gardeners)	231.52	212.39
WUA age	Age of the users' group	5.77	2.45
Original gardeners	No. of originally doing gardening in the irrigation area before rehabilitation began	84.61	134.15
Training	1= if leaders received training in the past two years; 0 otherwise	0.69	0.46
Plot gini	Economic heterogeneity (inequality measured as gini coefficient of plot sizes)	0.38	0.13
Soc. interaction	(population density) Household per unit irrigation area	23.18	21.41
Community characteristics			
Heterogeneity index	Index of ethnic fragmentation in village *i*.	0.17	0.19
Villages	No. of villages using an irrigation scheme	5.31	2.94
Ethnic villages	1= if villages using the scheme (≥75%) are from the same ethnic group	0.75	0.43
H2o distr. regime	1= if water distribution arrangement is continuous flow; 0 otherwise	0.37	0.48
Water schedule	1= if water distribution schedule was designed by the WUA.	0.73	0.44
Conflicts	1= if conflicts occurred over water distr. ; 0 otherwise	0.23	0.42
Conflict*	Predicted value of conflicts	0.23	0.25
Landlords resist.	1= if original landlords resisted land redistribution; 0 otherwise	0.08	0.26
Profitability	Predicted values of profit function	14.23	0.68
Market Access	1=Presence periodic market in community; 0 otherwise	0.63	0.48
Dist. To market	Distance from community to nearby market (km)	1.47	2.52
Bad opportunity	1= bad opportunities for alternative income sources in community i; 0 otherwise	0.78	0.41
Institutional variables			
Fine	1= penalty for breaking rules is fine; 0 otherwise	0.55	0.50
Forfeiture	1= heaviest penalty for rule breaking is forfeiture of plots; 0 otherwise	0.17	0.38
Warn	1= only warn those who break rules	0.27	0.44
Rule conformance	Predicted value of water poaching	0.42	0.36
Distrib. rule	Predicted value of water distribution rule	0.36	0.29
External factors			
Wage rate	Local wage per hour	1144.23	564.01
NGO proj.	1= past experience with community development initiatives; 0 otherwise	0.67	0.47
Sociorg.	No. of social organizations other than WUA	3.77	0.89

6.2.1.1.3 Results and discussion: Distribution rule

Table 6.2 presents a Probit estimates of the choice of water distribution rule. The magnitude and the direction of the predicted coefficients are important in our interpretation of the results. The coefficient of the variable representing the number of villages using the scheme is very significant and negative, indicating that the probability of observing a rotational arrangement of water distribution is high when many communities use a scheme, as a means of ensuring equity and building trust among beneficiary communities. Water shortage is highly associated with a continuous flow arrangement of water distribution, a fundamental problem that underlies the efforts being made by project staff to promote a rotational system of water distribution. We find a higher probability of observing a continuous flow arrangement where the WUA is younger. This result is consistent with the perception that older WUAs will choose a rotational system of water allocation because they might have learnt more efficient ways of allocating water and better ways of monitoring and enforcing agreements (Dayton-Johnson, 1999, 2000b). The results predict high probability of observing a continuous flow arrangement in socio-culturally heterogeneous communities, perhaps due to a lack of trust and high transaction costs of monitoring to prevent water poaching. It may be difficult to reach agreements over the modalities for implementing the rotational arrangement when tribal differences are high. The results also suggest that communities with previous experience with social organizations (e.g., NGOs) are likely to adopt a continuous flow system. This is inconsistent with the suggestion that experience with local organizations will promote collaboration, negotiation and peer monitoring required for a successful rotational allocation regime. A possible explanation for this observation could be that such communities instead learn to choose less costly means of implementing water allocation: avoiding the degree of organization and coordination (of dates of opening water, field preparation, etc) required under the rotational system.

Rule conformance has a positive and significant effect on the choice of the water allocation rule, i.e., suggesting that groups in which members conform to rules choose a continuous flow arrangement. This is contrary to our a priori expectation. The relationship between the water allocation rule and rule conformance is further examined in section 6.3.5. There is also a high probability of observing a continuous flow arrangement where there have been "illegal" extensions to the schemes. The results of this estimation support the field observation that conflicts over water distribution are highly associated with the time rotational mechanism of water distribution, because of the tendency of others to divert water into their fields when it is not their turn. The size of the WUA has no statistically significant effect on the probability of observing a particular water distributions rule.

Table 6.2: Probit model predicting the choice of water distribution rule

Variable	Coef.	Robust S.E	P>|z|	Marginal Effects dy/dx^	S.E	P>|z|
Villages	-0.2610	0.0812	0.001	-0.0797	0.0250	0.001
Water shortages	1.1363	0.5546	0.041	0.3283^	0.1351	0.015
Age of WUA	-0.0512	0.0212	0.016	-0.0156	0.0062	0.012
Heterogeneity index	2.7731	1.4003	0.048	0.8463	0.4377	0.053
Social org.	0.9939	0.3428	0.004	0.3033	0.1026	0.003
Maintenance*	3.7963	2.7777	0.172	1.1586	0.8573	0.177
WUA size	0.0007	0.0013	0.618	0.0002	0.0004	0.624
Conflicts	-1.2145	0.5800	0.036	-0.2778^	0.1039	0.008
Size of scheme	-0.0201	0.0399	0.615	-0.0061	0.0120	0.610
Dam extension	0.9161	0.5388	0.089	0.2581^	0.1328	0.052
Rule conformance	0.8719	0.5385	0.105	0.2529	0.1339	0.070
Landlord resist	-0.9587	0.5314	0.071	-0.2689^	0.1426	0.059
Plot gini	-1.9905	2.1691	0.359	-0.6074	0.6480	0.349
Profitability*	0.4699	0.5822	0.422	0.1426	0.1684	0.397
Constant	-9.8501	7.6339	0.197			
No. observations	52					
Wald χ^2 (14)	25.53					
Prob. > χ^2	0.0297					
Pseudo R^2	0.4071					

^ (dy/dx) is for discrete change of dummy from zero to one
*Predicted value

Communities that perceive the schemes as reliable (quality of maintenance is positive) are more likely to adopt the continuous flow arrangement. This is because better maintenance leads to lower losses from seepage, leakages through broken canal slabs and control structures. The result is also consistent with our a priori expectation that the rotational mechanism for water distribution is more likely to be observed in communities in which land lords resisted land redistribution. The executives of the user groups may want to use the control they have over the dam water to exert their influence and authority over the schemes. In this sense, the WUAs may adopt a water allocation principle that enables them to control the flow of water to individual plots to force compliance of regulations. Inequality does not have a statistically significant effect on the choice of the water allocation rule. However, the direction of effect of this variable suggests the probability of observing the rotational arrangement where inequality in land distribution is high.

In conclusion, the regression results show that the number of villages sharing the use of the same scheme strongly influences the choice of a distribution rule; a rotational water distribution regime is observed where many villages share the use of the same scheme. Age of the WUA (experience in water management), conflicts and resistance to land redistribution also have higher marginal effects of influencing the probability of a community choosing the rotational system. On the other hand, the continuous flow arrangement for water distribution is a popular choice in ethnically differentiated communities, and in communities that experience frequent water shortages.

6.2.1.2   Distributive conflicts

Participation and observance of norms of the user association can be motivated by expected economic incentives. The economic benefits, however, have distributional implications that are critical to the success of communal action for sustainable management of the resource. In particular, conflicts are a common feature in common-pool resource management (McCarthy et al, 2002), and invariably so in small-scale irrigation schemes. When resource allocation is perceived to be unfair it can lead users to irresponsible practices which can derail cooperative effort and render institutions for the management of the irrigation schemes dysfunctional.

Conflicts within the WUAs studied mostly stem from resource (land or water) distribution, but are usually resolved peacefully via conflict resolution mechanisms within the associations or local communities (involving chiefs and elders). Indeed, the question of which household gets irrigation land for dry season gardening is a major issue that continues to be a focus of programs supporting irrigation development for poverty alleviation in the study area. Similarly, incidence of conflicts over water distribution, as revealed by interviews with farmers in the sampled communities, can affect incentives to cooperate in maintenance.

Land or water conflicts can be internal to the village or external (with other communities using the scheme). However, the majority (over 70%) of the reported cases of water conflicts were internal, with many relating to the abuse of the water distribution rule (water poaching, see section 6.2.2.2). The present arrangement in which the water right is tied to ownership of plots within the irrigation area is also a source of conflicts.

Conflicts resulting from resistance to land redistribution and inequality in the allocation of costs and benefits can also be an important determinant of the success of community-based management of irrigation schemes in the study area (Engel, 2003). Conflicts over land allocation and water distribution are, however, endogenous and themselves determined by several factors. In section 6.2.1.2.1 below, we attempt to empirically analyze the sources of distributional conflicts in the survey communities.

6.2.1.2.1 Econometric analysis of distributive conflicts

In this section we examine why conflicts over resource allocation emerge in some communities but not others. Following the Probit specification described in section 6.2.1.1.1 above, the probability of land conflict occurring in community i is specified as:

$$\Pr ob(LC_i = 1) = x_i'\theta + \upsilon_i \qquad (6.8)$$

where LC_i is a dummy variable indicating whether land conflict occurred in community i; x is a vector of exogenous variables; θ is a vector of parameters to be estimated; and υ is an error term assumed to be normally distributed with mean zero and constant variance.

Conflict over land allocation is assumed to be influenced by several factors, including profitability, lack of opportunities to earn an alternative income, group size (number of gardeners), size of scheme, ethnicity of villages using the scheme, number of original land owners, intra-village ethnic heterogeneity, inequality in plot distribution, distribution rule (predicted value of water distribution rule used as proxy), population density (a proxy for social interactions), and the resistance of landlords to land redistribution (see Engel, 2003). Profitability is also endogenous, so the predicted value is used. *A priori*, it is expected that conflicts over land will be highly prevalent in communities where irrigation is more profitable. Lack of alternative economic opportunities will increase pressure on (and demand for) irrigation land which can lead to more conflicts. Similarly, more land conflicts are expected where landlords resisted land redistribution as well as where plots are unequally distributed. Moreover, large group size and ethnic fragmentation are also expected to lead to frequent conflicts over resource allocation as the tendency for people of different sociocultural backgrounds to distrust one another is high. However, more and frequent interactions among group members are expected to promote harmony and reduce conflicts. Similarly, less conflict is expected where communities sharing the use of the same irrigation scheme are of the same ethnic background. Land conflicts are expected to be lower in relatively larger schemes. However, the potential consequence for frequent water shortages makes lager schemes potential grounds for conflicts over water allocation. The water distribution rule adopted can also be the cause of conflicts. For example, attempts to turn water to ones fields when it is not his or her turn of the rotation can result in conflicts. Table 6.1 (above) also defines the variables used in the analysis of distributive conflicts.

6.2.1.2.2 Results and discussion: Distributive conflicts

The Probit regression estimates of the determinants of land conflicts in the scheme areas are presented in Table 6.3. The results support the hypothesis that land conflicts are more likely to emerge where irrigation is profitable. The scramble to gain more land is high in communities where irrigation is very profitable, thus increasing the probability of conflict occurrence. The probability that conflicts occur over irrigation land is also high where landlords resisted the land redistribution policy. This is largely expected. On the other hand, greater social interactions reduce the likelihood of land conflicts occurrence, as frequent interactions result in cordial relationships that make it possible to amicably resolve disputes. The direction of effect of the intra-community ethnicity variable is inconsistent with our expectation. However, the direction of effect of ethnicity of villages on land conflicts is consistent with our hypothesis. The results support our hypotheses that land conflict is more likely to emerge where the size of the user group is large, but decreases as the scheme size increases. All things being equal, the larger the irrigation area the larger will be the mean plot sizes and the lower the probability of land conflict occurrence.

Table 6.3: Probit estimates of determinants of land conflicts

Variable	Coeff.	Robust S.E.	P >|Z|	Marginal Effects^ dy/dx^	S.E	P >| Z |
Profitability*	1.4723	0.6299	0.019	0.2949	0.1312	0.025
Social interaction	-0.1126	0.0565	0.046	-0.0226^	0.0110	0.040
Bad opportunity	0.2411	0.6163	0.696	0.0444	0.1047	0.671
Plot gini	0.3289	2.1264	0.878	0.0659	0.4304	0.878
Landlord resist	3.9211	1.5289	0.010	0.9227^	0.0528	0.000
No. gardeners	0.0080	0.0046	0.084	0.0016	0.0009	0.089
Scheme size	-0.0747	0.0720	0.300	-0.0149	0.0140	0.286
Ethnicity of villages	-0.9278	0.7565	0.220	-0.2364^	0.2269	0.297
No. of Original gardeners	-0.0071	0.0025	0.006	-0.0014	0.0005	0.011
Hetero. index	-3.2148	2.0125	0.110	-0.6439	0.4004	0.108
Water distribution rule*	-0.8447	1.0216	0.408	-0.1692	0.2148	0.431
Constant	-18.93	7.7807	0.015			
No. observations	52					
LR χ^2 (11)	20.51					
Prob. > χ^2	0.0389					
Pseudo R-sq.	0.3489					

^ (dy/dx) is discrete change of dummy variable from 0 to 1
*Predicted value

The directions of effect of the variables representing inequality in land distribution, lack of alternative income earning opportunities, schemes sizes, and water distribution rule are consistent with our expectation but are not statistically supported by our data set.

It can be deduced from the forgoing analysis that, in a scheme, when the user group is large, individual plot shares are likely to be smaller and resistance to land redistribution is much more likely to increase the probability of the occurrence of conflicts over land. Moreover, where irrigation is profitable, landlords are more likely to resist land redistribution in an effort to gain greater benefits, thus increasing the risk of conflict occurrence.

Re-specifying equation (6.8), we examine the probability of conflict occurrence over water distribution. The parameter estimates of a Probit model predicting the probability of allocation conflicts are presented in Table 6.4.

The results show that the probability of observing water conflicts in the irrigation communities is higher in larger schemes. In schemes with large irrigation land, water requirements may go beyond the reservoir capacity, and the hustle to get water to one's plot can result in conflicts. We also observe high probability of conflicts occurring over water where water shortages are frequent. In times of water scarcity, when it becomes necessary to limit the amount of land cultivated to ensure that the available irrigation water is enough to meet the crop water requirement, the question of which farmer should be allowed to farm and who should not often raises mistrust that also results in conflicts. The regression results also support the field observation that the likelihood of water conflicts is higher in schemes where there have been illegal extensions. Often, farmers in the illegally extended portions after being served for some time "on humanitarian grounds" insist on their rights to irrigation water. Refusal of the WUAs to open water to the illegally extended parts of the scheme results in conflicts. We also find the probability of water conflicts to be low where the water allocation regime is a continuous arrangement. Although this allocation regime often disadvantages farmers located at the tail end of the schemes, it limits water poaching which has been a source of rancor among water users. Market orientation tends to increase conflicts over water while water conflicts tend to be frequent in younger WUAs. Older WUAs tend to more stable and might have developed better ways of managing conflict situations.

Table 6.4: Probit model predicting conflicts over water allocation

Variables	Coeff.	Robust Std error	P>|z|	Marginal Effects dy/dx^	Std error	P>|z|
Profitability*	0.7564	0.5296	0.153	0.0968	0.0651	0.137
Water shortages	0.9771	0.5517	0.077	0.1244^	0.0733	0.090
Villages	-0.1410	0.1010	0.163	-0.0181	0.0119	0.131
WUA size	0.0008	0.0013	0.529	0.00011	0.00017	0.529
Size of scheme	0.0689	0.0396	0.082	0.0088	0.0060	0.141
Dam extension	0.8663	0.4824	0.073	0.1020^	0.0709	0.150
Continuous flow*	-5.0788	1.2657	0.000	-0.6505	0.2122	0.002
Market access	0.9993	0.6545	0.127	0.2184^	0.1813	0.228
Age of WUA	-0.0809	0.0962	0.401	-0.0104	0.0125	0.410
No. of observations	52					
Wald χ^2 (9)	22.39					
Prob. > χ^2	0.0077					
Pseudo R^2	0.4351					

^ (dy/dx) is for discrete change of dummy from zero to one
*Predicted value

We can therefore conclude that water scarcity in the face of rising demand (extension of irrigation area) increases the likelihood of conflicts. Disputes over water are more prevalent in the rotational flow due to the tendency of others to illegally open their lateral gates for water to flow through when it is not their turn.

As the foregoing illustrations show, communities are faced with a variety of conflicts; internal and external, that can derail cooperation for collective management of the common-pool resource. Swift resolution of conflicts is important to improve confidence in the community of users, and help to build social cohesion required for effective collective action. The facilitating roles of NGOs and project management become very important.

6.2.2 Rule conformance

6.2.2.1 The supply of institutions

The success of collective action depends on the ability of the WUA to make and enforce institutions[37] that will govern the collective action. As has been mentioned before, irrigation water is a collective good, as the benefits of an improvement to it, achieved through maintenance, cannot be easily withheld from individuals who do participate in the maintenance activities. To overcome the problem of free-riding, therefore, requires that user groups design, through collective agreements, enforceable principles for sanctioning infractions of rules and rewarding compliance.

Based on the guidelines provided by LACOSREP, the user groups in the survey schemes have developed bylaws, which among others, include rules that protect the catchment area, maintenance obligations, levies (for gardeners, fishermen and livestock owners), tenure of office holders as well as meeting schedules. Generally, the bylaws not only stipulate obligations but also enable users to recognize their rights to the common good to generate confidence in cooperation. Rule enforcement, however, requires a credible threat of sanctions, if rules are not complied with, or the offer of positive incentives for compliance with rules. Prominent among the bylaws of the associations are sanctions for rule violation (e.g., failure to participate in collective maintenance activities, violation of the water allocation rule, non-payment of levies). The sanctions graduate from mere warning, through fines to forfeiture of ones right to farm at the schemes. In many cases, the amount of the fine is just equivalent to the prevailing local daily wages, but in others a multiple of this is charged as a punishment for rule violation.

Drafting of rules is one thing but rule enforcement is another. The ability of the WUAs to enforce their bylaws depends on whether the bylaws are legally recognized and for that matter on the authority to enforce rules. The source of authority could be legal recognition by the state, in which case there could be legal backing from state institutions. The WUAs, at the time of the survey, were in the process of being registered as cooperative organizations, a process which when ended can earn the users associations a legal recognition under Ghana's cooperative laws. But none of the WUAs had legal backing as at the time of the survey; in many cases they depended on the traditional authorities and moral norms to enforce rules. Gradual erosion and usurpation of the powers of the local authority make traditional institutions lose the potency to give legal backing required for long-term sustainability of the schemes.

However, internal mechanisms for enforcing norms (rules, bylaws, etc.) that govern collective behavior including penalties for noncompliance can affect

---

[37] The term institution as used here is defined to include all norms, values, rules and bylaws, as well as administrative arrangements that define the basis on which the schemes are managed.

the success of collective action for management of irrigation schemes. In the following section, we attempt to examine the factors that influence compliance with the rules and bylaws of the user groups to throw some light on the level of cooperation that prevails in the user groups for collective management of the schemes.

6.2.2.2   Determinants of Rule Conformance

6.2.2.2.1 Econometric model

Rule conformance is used here to depict the level of cooperation that exists in the user groups for collective management of the community irrigation schemes in northern Ghana. As determinants of cooperation for successful collection action, we examine two forms of rule conformance that manifest cooperation or lack of it. The forms of rule conformance used here to proxy cooperative behavior are payment of levies and water poaching. The presumption is that there is cooperation if members pay their levies promptly and the incidence of water poaching (i.e., stealing or directing water to ones field when it is not his turn, indicating the extent of violation of the water distribution rule) is a signal of the lack of cooperation.

To empirically analyze the determinants of rule conformance we specify the following reduced form specification of equation (5.14) in section 5.5:

$$Rule\ conformance = x'\delta + \varepsilon \tag{6.9}$$

where δ is a vector of parameters, x is a set of explanatory variables described below and ε is the error term assumed to be normally distributed with mean zero and a constant variance.

Levy payment and water poaching are measured as dummy variables in which the dependent variable takes the value of 1, for example, if members frequently pay their levies and the value of 0 if otherwise.

Following the Probit specification derived in section 6.2.1.1.1, we analyze the determinants of rule conformance by specifying two Probit models. The first is to examine the probability of levy payment (as a sign of cooperation) using the probit specification;

$$\Pr(Levy\ payment = 1) = x_i'\delta + \varepsilon_i \tag{6.10}$$

In the second, the probability of water poaching (violation of the water distribution rule) is used to examine conditions for non-cooperation, which can be detrimental to the success of collective action. The probit specification for water poaching is,

$$\Pr(Water \ poaching = 1) = x_i'\delta + \varepsilon_i \qquad\qquad (6.11)$$

Rule conformance is hypothesized to be affected by different factors including those relating to the community, the user group, the resource system, profitability, institutional and other external factors. In particular, variables featured in the analysis of rule conformance include the number of villages sharing use of the schemes, sanctions against rule violation, size of user group, water supply condition, conflicts over water, market access, experience with social organizations (e.g., NGOs), ethnic heterogeneity, inequality in plot allocation, resistance of landlords to land redistribution, profitability, techniques for rehabilitation of the scheme, water allocation rule, and ownership of rules (rules crafted by the user group). Conflicts over water allocation, water distribution arrangements, profitability and maintenance are endogenous to the system, as described in section 4.5. The predicted values of conflicts, profits and the water distribution regime are used. A two-stage estimation method described in Maddala (1983) for simultaneous equations in which one of the endogenous variables (e.g., maintenance) is continuous and the other endogenous variable (e.g. water poaching) is dichotomous is used to analyze the endogenous relationships in quality of maintenance (Table D6.2) and rule conformance. The following section presents comments on the variables used in the analyses of the alternative measures of collective action; rule conformance and quality of maintenance. Variable definition and descriptive statistics are as presented in Table 6.1 above.

6.2.2.2.2 Variables and hypotheses of factors affecting collective action on irrigation management at the community level

In this section we provide some comments on the hypothesized factors explaining rule conformance and the differences in the quality of maintenance of the irrigation systems presented in section 6.2.3. A long list of potential factors explaining collective action has been hypothesized (Berkes and Folke, 1998; Baland and Plateau, 1996; Ostrom, 1990; Wade, 1988), which need empirical justification. Many of the hypothesized determinants have been discussed extensively in chapter 5. However, the size of our data set cannot allow the empirical testing of all the factors discussed. Our analysis is limited to the hypothesized factors that are relevant to the case of northern Ghana. Some local specific factors that hypothetically explain the differences in outcomes of community-based irrigation management in the study area have also been included.

The organizational ability of the WUA is greatly affected by its size. The presumption in the literature is that cooperative behavior emerges easily when the size of the user group is smaller. This is because large groups are more difficult to organize, transaction costs of making and enforcing agreements is higher, and the incentive to cheat is high. Thus, as group size increases the benefit of cooperation decreases whilst the benefit of free riding on the cooperative effort of others increases. Larger WUAs on the other hand provide positive economies of scale in size; when it comes to pooling resources for maintenance as well as in self-monitoring of norms (Bardhan, 2000). Therefore, the effect of group size on the success of collective action is *ex ante* ambiguous.

The size of the scheme is likely to be an important determinant of cooperation. The cost of monitoring compliance with rules and bylaws would be higher in larger groups. Peer monitoring is expected to be easier in smaller irrigation schemes. Shared norms are common and social sanctions easier to implement in smaller schemes. We therefore hypothesize a negative relationship between scheme size and cooperative behavior.

The number of villages sharing the use of a scheme defines the extent to which a community's resources are shared with neighboring communities, which in turn can have implications for cooperation. The expectation is that user groups will be easier to organize when members reside in the same community. Coordination is expected to be easier and agreements easier to enforce when the number of villages using the scheme is smaller. This is because local norms of cooperation, which form the bases for monitoring and enforcement of bylaws in many rural communities, lose force when they cross boundaries (Dayton-Johnson, 2000a).

A credible threat of sanctions is expected to improve cooperation for successful local management of the irrigation systems. Threats of sanctions such as fines and forfeiture of plot are expected to increase rule conformance and participation in maintenance. Conversely, collective action is expected to be low where defaulters are only warned. Ownership of the rules and bylaws can also affect the level of compliance (Ostrom, 1990). We expect the rules designed by the user groups themselves to have greater acceptability and compliance.

Enforcement of rules is easier when population density is high, allowing for high social interaction among members. This is due to the fact that informal sanctions such as ridicule, gossip or ostracism are important means by which regulations are enforced in many traditional societies. Values that are shared in common by the group members, primarily those of honesty and respect, sustain trust and cooperative behavior in local communities. Social sanctions are easier to implement through reputation mechanisms which are easier to monitor if the extent of interactions among group members is high. A high population density is therefore expected to have a positive relationship with the quality of maintenance. Similarly, the level of cooperation is expected to be high in older WUAs. People get to know one another and their relationships improve as they interact for a longer period of time.

Direct economic benefits constitute one of the prime motivations for participation in collective action for the maintenance of the irrigation facility (White and Runge, 1992). Cooperation for collective action is therefore expected to be high in schemes where farmers grow crops that give them high economic returns.

A priori, water scarcity has an ambiguous effect on collective action. If water supply is abundant to the extent that no shortages occur there should be no incentive for farmers to participate in the maintenance of the irrigation schemes. In extreme scarcity and unreliability of water supply it is hard to achieve improvements through collective action. At moderate levels, however, an increase in scarcity (water shortage) is expected to have a positive effect on cooperative behavior and the quality of maintenance (Baland and Platteau, 1996; Tang, 1992)

Ethnic differentiation within a community is often hypothesized to have a negative effect on cooperation. The assumption is that it is often difficult for people of different ethnic backgrounds to agree and enforce social norms. On the other hand, social homogeneity may lead to stagnation of ideas, and can foster institutional inertia, thereby leading to lower overall institutional capacity as compared to communities with greater sociocultural diversity (McCarthy et al., 2002). Therefore, the impact of intra-community ethnic heterogeneity on cooperation is *a priori* ambiguous.

The ethnicity of villages sharing the use of an irrigation scheme can also affect the level of cooperation. Differences in social norms make creating and enforcing decisions across villages more costly (Dayton-Johnson, 2000a). It is expected that the quality of maintenance will be lower in schemes used by communities of different ethnic backgrounds.

The direction of effect of economic inequality is *ex ante* not clear (see for instance Agrawal, 2001; Baland and Plateau, 1999; Bardhan and Dayton-Johnson, 1999), as there are counteracting effects. Pitman (1995) suggests that groups involving people of similar status and power are more effective in generating trust and cooperation than more heterogeneous ones. On the other hand, in economically heterogeneous groups the more endowed members may find it in their interest to act as leaders and engage in more maintenance (Wade, 1997; Baland and Plateau, 1993; Olson 1965). We measure economic heterogeneity in terms of differences in landholdings at the irrigation schemes, measured as a gini coefficient of inequality in landholdings.

The effect of market access on the success of local management of irrigation is *a priori* mixed. On the one hand, better access to markets may increase the value of the irrigated crops, and thus the returns of managing the schemes efficiently, thereby favoring collective action (Agrawal and Yadana, 2001). On the other hand, better access to markets may tend to undermine the incentive to cooperate in a collective action by increasing the opportunity cost of labor, or by offering exit options that can make the enforcement of collective action norms difficult (Pender and Scherr, 1999, Baland and Platteau, 1996).

Two techniques were adopted in the rehabilitation of the dams studied: labor intensive (relying mainly on labor from the beneficiary communities and technical expertise of GIDA), and capital intensive (where work was done by contractors using heavy equipments). Where dam rehabilitation/construction was done by labor-intensive methods, the local communities develop a sense of ownership of the facility as the practice confers symbolic transfer of ownership. The learning effect of the labor-intensive technique improves the skills and capabilities of the WUAs to undertake quality maintenance. Apart from poor works done by many of the contractors engaged in the program, the capital-intensive strategy can result in a 'charity – recipient' mentality on the part of the beneficiaries. It is therefore hypothesized that the labor-intensive technique leads to higher-quality maintenance.

Higher outside wages result in higher opportunity costs of cooperation. This is especially so where there are more rewarding opportunities outside irrigation agriculture. The local wage is expected to have a negative effect on maintenance performance.

Cooperation may be habit forming, therefore, past experiences of cooperative engagements in the communities can be an important factor explaining the success of collective action. Where NGO activities and social organizations exist, communities are better endowed with social mobilization skills and norms to organize successful collective action for operation and maintenance of the schemes (Fujita et al, 1996). However, the effect can be negative if the activities of the NGOs tend to replace communal action in the villages. The effect of this variable on the success of collective action is therefore ambiguous, *a priori*.

6.2.2.2.3 Estimation results of rule conformance models

Table 6.5 presents Probit estimates of the probability that water levies will be paid. All the variables, but one, bear the expected signs. The results of an alternative specification in the Tobit framework which examines levy collection performance are presented in Table D6.3. The results show that enforcement of rules is important to induce cooperation for collective actions. Each of the variables representing fines and forfeiture of plots has a positive and highly significant influence on the level of compliance with water levy payment. These results support our hypotheses that strict enforcement of rules and norms established by the WUAs can increase the probability of cooperation for collective action. The marginal effect of enforcing bylaws (implementing, say, the forfeiture rule) increases the probability of compliance by about 0.7. The converse is that non-payment of water levies is high in communities where defaulters are only warned.

The predicted value of quality of maintenance has a positive and significant effect on rule conformance. Improvements in water supply conditions, as a result of quality maintenance, increase the probability of levy payment as people become content with the service provision. WUA size has a negative and significant impact on levy payment. This is also consistent with our a priori expectation that increasing WUA size leads to a decrease in the probability of cooperation. Similar results suggesting an increased likelihood of success of smaller groups have been reported for user governed forestry institutions (see Cernea, 1989). The direction of the effect of the coefficient of the variable representing water shortage is consistent with our *a priori* expectation. It is negative and highly significant indicating that water shortages lower the probability of compliance with rules for levy payment. As noted above, the water supply condition constitutes one of the important factors that influence attitude towards fee payment. People will be less willing to pay if the water supply condition cannot be reliable. Frequent water shortages decrease the probability of cooperation.

Table 6.5: Probit estimates of rule conformance (Levy payment)

Variable	Coeff.	Robust S.E	P>|z|	Marginal Effects Coeff.	S.E	P>|z|
Fine	1.9040	1.2542	0.004	0.6577^	0.1567	0.000
Forfeit	2.7599	1.2542	0.008	0.6593^	0.1354	0.000
Maintenance*	0.0012	0.0004	0.000	0.0005	0.0001	0.000
WUA size	-0.0131	0.0039	0.001	-0.0052	0.0016	0.001
Water shortages	-1.0112	0.5705	0.076	-0.3862^	0.1983	0.051
Conflict*	2.5238	0.8816	0.004	0.5346^	0.0968	0.000
Market access	-0.7813	0.5229	0.135	-0.3010^	0.11897	0.113
Social org.	1.9032	0.8844	0.031	0.5720^	0.1511	0.000
Hetero. index	-1.2329	1.8794	0.512	-0.4915	0.7499	0.512
Plot gini	-2.5124	2.2978	0.274	-1.0015	0.9173	0.275
Soc. interaction	0.0939	0.0397	0.018	0.0374	0.0159	0.018
Landlord resist	-5.6202	1.9079	0.003	-0.6814	0.0882	0.000
Lab. intensive	0.9219	0.7168	0.198	0.3538	0.2556	0.166
Profitability*	0.8603	0.5296	0.104	0.3429	0.2113	0.105
Water allocation rule*	0.3385	0.8086	0.675	0.1349	0.3225	0.676
Constant	-13.5844	7.4515	0.068			
No. of observations	52					
Wald $\chi^2(15)$	22.53					
Prob. > χ^2	0.0947					
Pseudo R^2	0.4387					

(^) dy/dx is for discrete change from dummy variable from 0 to 1
*Predicted values of variables

Contrary to our *a priori* expectation, the conflict variable shows a positive and significant effect on rule conformance. This is unexpected since one would have expected that people would be reluctant to pay levies where conflicts are prevalent and people could not trust one another. However, as has been argued earlier, in a conflict zone, such as the case of many of the survey communities, where NGOs and religious bodies continue to preach peace and harmony, acts that encourage group cohesion are highly encouraged.

The coefficient of the variable for market access has a negative effect on levy payment, but the variable is statistically significant only at the 13% level. The results show that compliance with the water levy payment rule is low where markets are accessible. Thus, the direction of effect of this coefficient is consistent with the argument that market access has an adverse impact on cooperative behavior, as it generates alternative interests that can undermine institutions. The variable denoting community experience with social organizations is positive and significant. This suggests that communities with prior experience with NGO supported projects are more experienced in community mobilization and fee collection which is often a condition for external support. This increases the probability of cooperation in collective action.

Population density at the schemes, a proxy for social interactions, has a positive and significant effect, indicating that high social interaction increases the probability of levy payment. Peer monitoring may be easier when there is closer interaction among gardeners. Knowledge of the fact that others have paid their levies can motivate compliance with the levy payment order.

The resistance of landlords to the land redistribution policy has a negative and significant effect on levy payment. This implies that the probability of cooperation is lower in communities where landlords resisted land redistribution. There is divided allegiance where landlords allocate plots; this generates factionalism and shifting alliances that can be a barrier to cooperative behavior. Those who obtain irrigable land from landlords may not feel obliged to pay water levies if the water distribution mechanism cannot be used to force compliance. Profitability has an increasing effect on the probability of levy payments, but the variable is only marginally significant at the 10% level. The variables representing the labor intensive mode of rehabilitation and the water allocation rule are positive in their effect but not statistically significant.

The coefficient for the inequality term is negative but statistically not significant. The direction of effect is, however, consistent with the results of other studies (see for instance Lam, 1994). A high level of inequality leads to low levy payment. The intra-community ethnic fragmentation term is also negative but not statistically significant. Thus, suggesting that heterogeneity based on social or economic factors can raise suspicion, lower trust and make it difficult for people to cooperate. Accountability of leaders was dropped from the list of independent variables because it perfectly predicted the dependent variable.

We now analyze the factors that encourage non-cooperative behavior and for that matter free riding in the local management of irrigation schemes in northern Ghana. Thus, Table 6.6 presents Probit estimates of the frequent violation of the water allocation rule, the opposite of cooperation. In particular, we examine the underlying determinants of the frequency of violation of water allocation rules.

The variable representing the number of villages sharing the use of the same irrigation scheme has a positive and significant effect on water poaching. This indicates that the probability of water allocation rule violation is high in schemes that are shared by many villages. Apart from the high costs of monitoring compliance across villages, social differentiation lowers cooperation.

This is because social norms are difficult to enforce across communities. The results, therefore, support our assumption that cooperation is less likely to emerge in schemes that involve multiple villages (see also Meizen-Dick et al, 2002)

Table 6.6: Probit Estimates of Violation of Water Allocation Rule

Variable	Coeff.	Robust S.E	P>|z|	Marginal Effects Coeff.	S.E	P>|z|
Villages	0.2483	0.0909	0.006	0.0929	0.0348	0.008
Water shortages	-0.6720	0.4804	0.162	-0.2415^	0.1688	0.152
Landlord resist	3.2324	1.1826	0.006	0.7233^	0.0772	0.000
H$_2$O alloc rule by WUA	-1.9049	0.7107	0.007	-0.6574^	0.1735	0.000
Fine	-1.3361	0.7683	0.082	- 0.4794^	0.2497	0.055
Forfeit	-2.6281	0.8533	0.002	-0.5333^	0.1033	0.000
Maintenance	-2.3137	2.8059	0.410	-0.8660	1.0517	0.410
Plot gini	6.8567	2.2288	0.002	2.5665	0.8394	0.002
Labor intensive	-3.1129	0.7083	0.000	-0.8279^	0.1041	0.000
Age of WUA	0.5138	0.1224	0.000	0.1923	0.0486	0.000
Soc interaction	-0.0120	0.0149	0.421	-0.0045	0.0056	0.418
Water allocation rule*	-2.0839	1.1664	0.074	-0.7800	0.4332	0.072
Profitability*	-1.0619	0.5894	0.072	-0.3975	0.2284	0.082
Scheme size	0.1206	0.0449	0.007	0.0451	0.0177	0.011
Market access	0.7564	0.5370	0.159	0.2666	0.1765	0.131
Constant	-11.7134	7.6164	0.124			
No. observations	52					
Wald χ^2 (15)	31.34					
Prob. > χ^2	0.0079					
Pseudo R^2	0.4791					

(^) dy/dx is for discrete change from dummy variable from 0 to 1
*Predicted values of variables

Consistent with our *a priori* expectation, water poaching is positively correlated with group size. Thus, indicating that probability of rule violation is higher when the user group is large. This supports the conclusions of a theoretical work of Weissing and Ostrom (1990), which shows that in equilibrium an increase in the number of irrigators is associated with an increase in water theft, *ceteris paribus*. Monitoring is difficult and rule enforcement is weak in larger groups. Water poaching can, however, be minimized when self-monitoring by all group members is encouraged.

The results also show that the variable representing water shortage has a negative and significant effect on the probability of violation of the water allocation rule. Inconsistent with our *a priori* expectation, the result shows that water shortages decrease the probability of water rule violation. We could, however, reason that frequent water shortages increase vigilance on the part of users and thus promote peer monitoring, which could result in lower levels of water theft.

Consistent with the previous finding, the variable for landlord resistance to land redistribution has a positive coefficient which is highly significant. This implies that, given the other factors, resistance of landlords to land redistribution increases the probability of water theft. Indeed, in those communities the legitimacy of the WUA is greatly undermined and rule enforcement mechanisms are weak.

The probability of rule violation is low in user groups that developed their own distribution rules. The coefficient of the variable that indicates that the water allocation rule was developed by the WUA has a negative and significant effect on the probability of water stealing. This implies that, there is a decreased probability of water stealing when the water allocation rule is crafted by the communities themselves. This finding matches Ostrom's (1990) observation that rules devised by user groups themselves have better chances of acceptability and durability than externally imposed institutions.

The institutional variables, fine and forfeiture of plots, have a significant reduction effect on the incidence of water poaching. These results reinforce our assumption that the probability of water allocation rule violation is lower in communities where sanctions are seen to have biting effect. Stiffer penalties for rule violation have consistently shown a reduction effect on the probability of rule violation.

Inequality in landholding at the schemes is positive and significant. This indicates that economic inequality promotes rule violation. This supports our *a priori* assumption that cooperation in collective action will be difficult to achieve in groups that are internally differentiated in terms of resource endowment. It also suggests that rules are probably violated more often by well-endowed farmers, because they can get away with such violations. Bardhan, (2000) and Dayton-Johnson (2000) observe a similar negative effect of inequality on cooperation for collective action in their studies in India and Mexico, respectively.

The variable representing the labor intensive technique of dam rehabilitation has a highly negative and significant effect on the probability of water allocation rule violation. This implies that water theft is significantly lower in communities where the schemes were rehabilitated by the labor intensive technique. The technique promotes good relationships among participants as they work together to achieve a common goal. Mutual commitment to ensure the sustainability of the schemes encourages cooperation for the success of collective action.

The variable representing age of a WUA also has a positive and significant effect on the probability of water rule violation. This is inconsistent with our assumption that the level of cooperation increases with the age of a WUA. This result, however, suggests that people find means to get away with laid down rules as they become more familiar with the system and its institutions. The probability of water theft is thus high in older WUAs than new ones. The coefficient of population density is negative and statistically significant, indicating that higher social interaction, made possible by high population density at the schemes, decreases the probability of rule violation.

Rule violation is easier to detect where population density is high and there is high interaction among irrigators. The violation of water allocation rules (water theft) is also influenced by the water distribution rule. The sign of the predicted continuous flow rule variable is negative and significant. This implies that the higher the probability of observing continuous flow regime the lower there is water poaching. This supports our earlier assertion that water theft is more frequent in WUAs with the time rotation mechanism of water distribution.

Profitability has a negative and significant effect on the probability of water allocation rule violation. This is contrary to our *a priori* expectation. However, the urge to protect economic interests arising from higher profits can motivate peer monitoring to reduce the chances of water stealing. Thus, the higher the expected return the greater the motivation to engage in self-monitoring to prevent water poaching. Scheme size has a statistically significant effect on the violation of the water allocation rule. This is consistent with the hypothesis that the probability of water poaching increases with the size of the irrigation area. This is because the costs of monitoring are high and rule violations are difficult to detect in larger schemes.

The sign of the variable representing market access suggests that market orientation increases the probability of water rules being violated. This variable is however only statistically significant in its effect on rule violation at the 13% level.

The robustness of the findings presented in this section is supported by the general consistency in the directions of the effects portrayed by the variables estimated in the two models for rule conformance. The majority of the variables are in conformity with our a priori expectations.

6.2.3 Quality of maintenance of community irrigation schemes

Maintenance is an essential component of irrigation systems management, as it governs their long term sustainability. Therefore, the success of any community-based irrigation management will, among others, depend on the ability of users to provide the collective action needed to improve the performance of the system. The level of maintenance provided at the schemes illustrates the capacity of local communities to maintain the systems. In this section, the quality of maintenance (i.e., the state) of the irrigation schemes is used as a measure of success of cooperation for collective action.[38]

---

[38] Total resource contribution could have been used. However, as Meizen-Dick et al. (2002) aptly clarify that the level of resource mobilization is affected by several factors. Physical condition and periodicity play important roles. Groups that do not maintain their schemes might have to mobilize more resources for maintenance at a time, whereas the ones that had been maintaining the scheme regularly might have a lower level of resource mobilization because they have been performing the maintenance task frequently.

6.2.3.1   Maintenance Performance

A fundamental task of the WUA is the mobilization of resources for regular and timely maintenance of the irrigation system. Maintenance activities carried out the by WUA and the extent of household participation in maintenance schedules have been examined in section 4.4. In this section we analyze the quality of maintenance achieved by the user group.

Whether or not the user group cooperates in the maintenance of the irrigation schemes could not be observed directly, but generally the extent of cooperation in the maintenance of the community schemes manifests itself in the quality of maintenance the community produces (see McCarthy et al, 2003; Bardhan, 2000; Dayton-Johnson, 2000a), which invariably depends on the ability of the WUA to mobilize labor and other resources for maintenance activities. Thus, a principal indicator of a group's success in mobilizing resources for maintenance is the state of repair of the irrigation infrastructure.

In order to determine the state of the infrastructure, an irrigation system inspection was conducted at each of the schemes surveyed as described in section (4.2). An index of quality of maintenance of the irrigation systems was constructed from a set of indicators collected during the system inspection, covering: (1) the state of the dam infrastructure (cracks in dam wall, seepage at dam toe, rip-rap on upstream slope, erosion on down stream slope, structure of spillway, inlet/outlet structures, control valves, etc), (2) the state of canals and drains (cleanliness of canals, drains, and laterals; state of slabs, etc), (3) the state of the catchment area (vegetative cover, bunds, grass cover on bunds, etc), and (4) other measures undertaken by the WUA to protect the irrigation system. The state of each of the elements was ranked on a scale of 0 – 5 (very poor to excellent). Table 6.7 presents a summary of the values of these categorical variables.

Table 6.7: Distribution (percent) of key indicators of state of maintenance of the irrigation infrastructure

Condition of ..*	V. Poor	Poor	Fair	Good	V. Good	Excellent
Dam embankment	0	4.5	41.0	28.2	24.4	1.9
Side slopes	1.9	2.5	34.6	36.6	24.4	0
Spillway	0	6.7	25.9	44.3	23.1	0
Physical structure of canal	7.7	4.8	42.3	39.5	5.7	0
Cleanliness of canals	6.7	5.8	46.2	31.7	9.6	0
Secondary canals	1.9	31.7	50.9	9.7	5.8	0
Main drain	0	29.8	58.7	11.5	0	0
Catchment area	0	5.8	28.8	51.9	13.5	0
Bunds	25.9	22.2	24.0	25.0	2.9	0

Sample size = 52.

*These measures are part of a larger set of indicators of the quality of the irrigation system

From the categorical variables, an index depicting the state of the irrigation infrastructure (a measure of quality of maintenance) is estimated. Following

Asfaw (2003), an irrigation system's quality index for scheme j was calculated using the following relationship.

$$Qindex_j = \sum_{i=1}^{n} (Score_i - Minimum\ score) \Big/ (Maximum\ score - Minimum\ score) \qquad (6.12)$$

where Qindex refers to quality index, Score = score of community j; $n(=23)$ is the number of components (indicators) evaluated. The minimum score is 23 and the maximum score is 92.

The quality index scores calculated are not only low but show differences in maintenance performance across schemes as depicted in Figure 6.4 below. The trend portrayed in the figure underscores the fact that the willingness and ability of user groups to undertake sustainable maintenance of the schemes vary, thus taking exception of the often held simplistic and optimistic views of the potential of local organizations in resource management.

Figure 6.1: Trends in quality of maintenance across communities

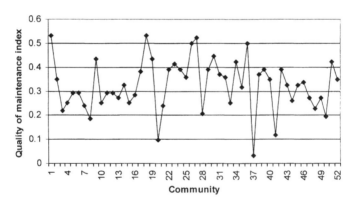

Given that the sustainability of the schemes depends on the quality of maintenance it is essential to examine the factors that account for the differences and the conditions for successful collective action for sustainable maintenance of the community irrigation systems.

6.2.3.2 Determinants of quality of maintenance of the irrigation schemes

6.2.3.2.1 Econometric model specification

Quality of maintenance is affected by several factors with associated structural relationships (interdependencies) as outlined in section 5.5. Following Bardhan (2000) and Dayton-Johnson (2000a) the observed quality of maintenance was used as an indicator of cooperation for collective action in the management of the community irrigation schemes. Based on this indicator, a model specifying the quality of maintenance was used to econometrically analyze the determinants of success of collective action for the management of the local irrigation systems.

Recalling from our conceptual model, a reduced form of the structural maintenance equation is estimated. In particular, quality of maintenance is presented as:

$$Ma\text{int}enance = x'\beta + \ell \qquad (6.13)$$

where x_i is a vector of explanatory variables including theoretically hypothesized as well as community specific factors that determine the success of collective action for the management of local commons; ℓ_i is a disturbance term assumed to be normally distributed with mean zero and constant variance.

In an attempt to empirically explain why some communities perform better than others, in terms of maintenance of the schemes, we specify the index of the quality of maintenance function in an ordered response framework. From the quality index, a three-level ordered response ("1" bad, "2" fair and "3" good) was also constructed. Letting y denote the observed quality of maintenance in the ordered response taking the values (J=1, 2, 3), we can derive an ordered Probit model for y from a latent variable model. Assume a latent variable y* is determined by:

$$y_i^* = x_i'\beta + \ell_i, \qquad (6.14)$$

for community i=1, 2,..,n, where x_i is the vector of explanatory variables, β is the vector of parameters, ℓ_i is a disturbance term assumed to be normally distributed with mean zero and constant variance (i.e., $\ell_i \sim NID(0, \sigma_i)$). Let $b_1 < b_2 < b_3$ be unknown cut points (also referred to as threshold parameters) to be estimated. Then, the ordered Probit model with J alternatives is defined as follows:

$$y_i = \begin{cases} 1 & if \ y_i^* < b_1 \\ 2 & if \ b_1 < y_i^* \le b_2 \\ 3 & if \ y_i^* \ge b_2 \end{cases} \qquad (6.15)$$

For example, y taking on the values 1, 2, and 3 imply there will be two cut points b_1 and b_2. Given the standard normal assumption about ℓ_i each response probability is computed as:

$$P(y_i = 1) = P(y_i^* < b_1) = P(x_i'\beta + \ell_i < b_1) = \Phi(b_1 - x_i'\beta)$$
$$P(y_i = 2) = P(b_1 < y_i^* \leq b_2) = \Phi(b_2 - x_i'\beta) - \Phi(b_1 - x_i'\beta) \qquad (6.16)$$
$$P(y_i = 3) = P(y_i^* \geq b_2) = 1 - \Phi(b_2 - x_i'\beta)$$

where Φ is the cumulative distribution function for ℓ_i and is assumed to contain no additional unknown parameters, so that, for example ℓ_i has a known variance. This assumption helps to fix the scale of the measurement of y_i^* but not the origin. Identification is achieved by assuming that x_i does not contain a constant term. In general, equation (6.16) can be compacted as

$$P(y_i = j) = \Phi(b_j - x_j'\beta) - \Phi(b_{j-1} - x_j'\beta) \qquad (6.17)$$

The coefficients b and β can be estimated by maximum likelihood. For each community j, the log likelihood function can be stated as:

$$LogL = \sum_{i=1}^{n}\sum_{j=1}^{J} y_{ij} Log\left[\Phi(b_j - x_i'\beta) - \Phi(b_{j-1} - x_i'\beta)\right] \qquad (6.18)$$

where

$$y_{ij} = \begin{cases} 1 & \text{if } y_i = j \\ 0 & \text{else} \end{cases}$$

Maximized with respect to ($\beta, b_1, ..., b_{J-1}$) with $M + J - 1$ parameters, where M = number of exogenous variables. (β (and hence M) does not include an intercept.) Thus, the ordered model for the analysis of the quality of maintenance of the irrigation systems is presented as:

$$Oprobit\,(ordered\ index)_i = \beta_1 Dist_i + \beta_2 GC_i + \beta_3 CC_i + \beta_4 AR_i + \beta_5 \Pi \atop + \beta_6 Inst_i + \beta_7 Ext_i + \ell_i \qquad (6.19)$$

Variables are as defined above. The results of an alternative Ordinary Least Squares specification, using the quality of maintenance index as the dependent variable is presented in Table D6.2.

Theoretically, factors ranging from the characteristics of the resource system, user group characteristics, to institutional as well as external factors have been hypothesized to affect the ability of local organizations to organize successful collective action for the maintenance of community irrigation schemes. As described in section 6.2.2.2.2, the characteristics of the resource include variables such as scheme size, water supply condition, quality of the rehabilitation work, rehabilitation technique, population density at the scheme and profitability. Group and community characteristics include WUA size, inequality, training of WUA executives, rule conformance, intra-community ethnic heterogeneity, ethnicity of villages sharing the scheme, and landlords resisting land redistribution. The institutional variables used in the analyses are sanction regimes that include fines and forfeiture of rights to plots, while market access and local wage are some of the external factor that can affect collective action on commons management. However, as illustrated in our conceptual model, profitability, allocation rule and rule conformance are endogenous to the system. Predicted values of these variables generated from first-stage estimations are used in the analysis presented in this section. In particular, the predicted values for rule violation (water poaching in Table 6.6) is used as the proxy for rule conformance while the predicted values for continuous flow mechanism of water allocation (Table 6.2) represent the distribution rule.

6.2.3.2.2 Results and discussion: Quality of maintenance

The empirical results of the analysis of the determinants of quality of maintenance, a measure of the level of collective action achieved in the management of community-based irrigation schemes in northern Ghana, are presented below.

Table 6.8 presents the results of the ordered Probit estimation. All the variables but two are consistent with our *a priori* expectations. The lower and the upper thresholds are both statistically different from zero. Both the magnitudes and signs of the regression coefficients are of major importance in our interpretation of the results presented in this section.

The group size has a negative and statistically significant effect on the quality of maintenance. This implies that quality of maintenance is lower in larger groups. This supports our hypothesis that quality of maintenance declines with group size, which is also consistent with Olson's (1965) group size hypothesis. As emphasized in the preceding sections, cooperation for successful local management is more likely to emerge and be sustained in smaller groups. This is because smaller group sizes present opportunities for frequent interactions and to build reputations while expectations about future interactions increase the value of reputations for cooperative behavior, mutual monitoring, and foster a high level of trust which create conditions that are conducive to collective action (Poteete and Ostrom 2000).

The results our analyses show that the quality of maintenance is high where villages sharing the use of the same scheme are of the same ethnic identity. This suggests that organizing across ethnically diverse communities can impose higher costs and consequently lead to a lower level of cooperation, *ceteris paribus*. Social norms are easier to enforce across villages of ethnic similarities. Similar findings have been reported in Dayton-Johnson (2000a) and Bardhan (2000) that suggest that cultural homogeneity enhances cooperation and reduces inter-community conflicts. Though not statistically significant, the coefficients of the variable representing economic inequality (inequality in land holding) indicate a positive effect of inequality on community maintenance performance. The coefficient of the variable for the use of labor intensive techniques in the maintenance programs also has a positive sign in accordance with our hypotheses, but the variable is not statistically significant in its effect on quality of maintenance.

Table 6.8: Ordered probit estimates of quality of maintenance of the irrigation schemes

Variable	Coefficient	Standard Error	P|z|	Marginal effect of outcome (Good = 3)		
				Coef.	S.E	P|z|
WUA size	-0.00283	0.00132	0.033	-0.00079	0.00039	0.043
Plot Gini	1.39046	1.81430	0.443	0.39195	0.52145	0.452
Training	2.56449	0.88171	0.004	0.27784*	0.08104	0.001
Forfeit plot	2.16354	0.82149	0.008	0.71683*	0.19523	0.000
Pay fine	2.13047	0.68714	0.002	0.52174*	0.14960	0.000
Water scarcity	1.20402	0.52216	0.021	0.32140*	0.13671	0.019
Labor intensive	0.02927	0.38921	0.966	0.00818*	0.19048	0.966
Ethnicity of villages	1.04568	0.61646	0.090	0.23079*	0.10329	0.025
Distance to market	0.00471	0.08731	0.957	0.00133	0.02461	0.957
Rule conformance°	-2.04842	0.82215	0.013	-0.57742	0.24234	0.017
Wages per hour	-0.00055	0.00417	0.189	-0.00015	0.00012	0.182
Quality of rehab. work	-0.12845	0.50044	0.797	-0.03538*	0.13429	0.792
Landlords resist	-2.30903	0.96008	0.016	-0.25445*	0.07849	0.001
Water allocation rule°	0.10817	0.76852	0.888	0.03049	0.21610	0.888
Profitability°	0.63727	0.45723	0.163	0.17964	0.13604	0.187
_Cut 1	11.68665	6.27336				
_Cut 2	13.16610	6.31698				
No. observations	52					
LR χ^2 (15)	38.86					
Prob. $> \chi^2$	0.0007					
Pseudo R^2	0.3403					

(*) dy/dx is for discrete change of dummy variable for 0 to 1
(°) Predicted values

The directions of effect of all the coefficients of the institutional variables are positive, and are statistically significant (see also Gyasi et al., 2004). They support our hypotheses that sanctions such as payment of fines or loss of opportunity to farm (forfeiture of plots) for failing to participate in collective maintenance activities promote greater maintenance participation and hence higher-quality maintenance. This backs existing results from field and experimental studies concluding that without effective institutions common-pool resources will not be sustainably managed (see for instance, Larson and Bromley, 1990; Ostrom et al., 1994; Tang, 1992; Bromley et al., 1992).

The results of the analysis also show training to have a positive and significant effect on the quality of maintenance. Quality of maintenance is high for schemes where leaders receive frequent training. Training improves the capabilities of the user groups to undertake quality maintenance. Continuous post-rehabilitation training support to the WUA can enhance group cohesiveness and cooperative behavior required for successful collective action. The coefficient of the variable representing water shortage shows a positive and statistically significant effect on the quality of maintenance of the schemes at the community level. This shows that communities that do not experience water shortages may not be serious with their maintenance schedules. Implicitly, water shortages induce the user groups into action, especially in communities with past experience of hardships as consequences of water scarcity that resulted from the collapse of the schemes.

The resistance of the original landowners to a land redistribution policy negatively affects the quality of maintenance of the schemes. The results support our initial assumption that quality maintenance is low in community schemes where landlords resisted the land redistribution policy. Apart from undermining the authority of the WUA, the noncooperative behavior of landlords could have a negative effect on group cohesion and unity required to organize successful collective action.

Perception about the quality of rehabilitation work done by contractors does have a significant effect on the quality of maintenance. But the direction of the effect is opposite to the impact that would be consistent with our a priori expectation. The result is also inconsistent with the effect of community perception about the quality of rehabilitation work on the household's decision to participate in collective maintenance activities (see section 6.3). This is a puzzling result since one would have expected that quality of maintenance would be poorer where the communities have cause to complain about the quality of the rehabilitation work, It, however, appears that the user groups have been able to improve the quality of the schemes, that were perceived to have been poorly rehabilitated, through collective action.

The coefficient of the local wages term showed a negative affect on the quality of maintenance of the irrigation schemes. The extent to which this variable affects household decisions to contribute efforts to the maintenance of the schemes is further explored in section 6.3. Although not statistically significant, distance to markets is positively related to maintenance quality, implying that increased access to markets leads to decreased quality of maintenance. As stated earlier, this is likely to be due to the fact that the increased exit options that become available as a result of market access may reduce reliance on the irrigation facility and for that matter reduce commitment towards the sustainability of the schemes. The low level of significance may, however, be due to the potential counteracting effects described earlier.

Profitability has a positive impact on quality of maintenance. We find that the quality of maintenance is higher in profitable schemes because of the economic interests at stake. This supports the assumption that the viability of the irrigation scheme is an important factor that conditions participation and success of cooperative management of the schemes. If farmers do not make enough money from irrigation they may not pay fees or contribute efforts towards maintenance, and if the WUAs cannot collect enough fees or mobilize enough labor they will not be able to carry out the operation and maintenance tasks efficiently.

The variable representing rule conformance has a significant effect on the quality of maintenance of the irrigation schemes. That is, frequent violation of rules governing the management of the schemes negatively affects the ability of user groups to achieve successful collective action for the maintenance of the community irrigation schemes. Therefore, quality of maintenance is poorer in schemes where rules are frequently violated. Water allocation rules on the other hand did not have a statistically significant effect on the quality of maintenance.

6.2.4 Section conclusion

A policy shift towards community participation in irrigation management has become widespread, but the success of these policies depends on our understanding of the factors that facilitate or constrain effective community participation. In this section, we have examined distribution rules as well as rule conformance, and implications of these and others factors suggested in the theoretical literature for the success of local management of irrigation systems in northern Ghana.

Overall, the estimated equations provide evidence that most of the hypothesized factors explain various aspects of collective action. The study finds that water distribution arrangements varied across schemes. Age of the water users association, number of villages sharing use of the irrigation scheme, conflicts, and the resistance of landlords to the land redistribution policy strongly favor the likelihood that a system chooses the rotational regime of water allocation. However, the water allocation arrangement chosen can also be a source of distributive conflicts.

The results show that collective action is more likely to succeed where the size of the user group is small, where villages using the schemes are few and of the same ethnic group, and where there is equity in the distribution of benefits (plot allocation is equitable). There is less cooperation (non-payment of levies, rampant water poaching and poor quality of maintenance) in schemes where landlords have resisted the policy to redistribute land among irrigators to promote equity. Profitability of the irrigation enterprise is essential for the success of local management of the systems: if farmers do no make enough money from irrigating, they will not pay fees or contribute times. If the WUA is unable to collect enough fees or mobilize labor it cannot effectively carry out operations and maintenance activities. Training was found to have a highly positive impact on the quality of maintenance. The role of facilitators and change agents is therefore crucial for successful collective action, as training promotes behavior that favors collective action and enhances capability of the managers of the user groups.

Rehabilitation techniques that enable user group participation lead to a better quality of operation and maintenance. The leaning effect of the labor intensive technique not only improves capabilities of the WUA to undertake quality maintenance but also promotes group cohesion, the sense of belonging and community identity with the physical infrastructure, and promotes moral commitment to ensure the sustainability of the schemes. Resistance of landlords to land redistribution had a negative effect on cooperation. It promotes rule violation and creates shifting alliances that affect the ability of the WUAs to organize successful collective action for the maintenance of the schemes.

It may be politically prudent to site the schemes in such a way so as to ensure that the benefits spread to many communities, but potentially at the expense of sustainability as the involvement of multiple villages in the use of the same irrigation schemes is not likely to produce the level of cooperation that is necessary for successful collective action in the maintenance of the systems. Decentralization of group activities to sectional levels where villages could come under different sections can promote cooperation. Market access has a negative effect on cooperation for collective maintenance. Better exit options made possible through market integration generate alternative interests, which can undermine institutions for the management of the schemes. Similarly higher wages outside irrigation increase the opportunity costs of labor and reduce the maintenance quality.

The results of this study also suggest that a credible threat of sanctions is important to ensure cooperation (rule conformance) and high quality of maintenance. Local rules and by-laws including fines and forfeiture of the right to benefit from the scheme, play very important roles in eliciting cooperation for collective management of the schemes. Rule conformance is high where rules were seen to be crafted by the WUA. Since institutions help to guide human behavior, clear rules regarding participation are important for the success of community-based management of the irrigation resource. It is essential that farmers enforce bylaws and regulations, strengthen monitoring and evaluation mechanisms to ensure efficient and sustainable management. Policy measures that facilitate the development of local institutions could enhance the ability of local organizations to sustainably manage their resources.

Overall, the regression results are consistent with the hypotheses regarding the factors that determine the success of collective action for the maintenance of community-based irrigation schemes in northern Ghana. Understanding the factors that influence household participation in the collective maintenance activities organized by the user groups will provide additional insights into how communities differ in terms of ensuring sustainable management of the irrigation facilities. The following section addresses this issue.

6.3 Household incentives for participation in collective action for irrigation maintenance in northern Ghana.

In this section, the analysis of collective action is taken further by identifying the factors determining household participation in the collective maintenance of the irrigation facilities.

As noted in section 5.5, households decide whether to contribute to the maintenance effort, the aggregate amount of which affects community maintenance performance. We have collected data on the amount of time each household spent on participation in maintenance work at the irrigation scheme over 12 months preceding the survey period. The data shows that the hours of work range from $0 - 151$ (C.V. $= 0.89$). The variance in labor contribution is sufficiently large to suggest the tendency for some households to renege on their obligations to the maintenance of the scheme. Using an agricultural household model, we attempt to empirically examine the factors that influence household decision to contribute efforts towards the maintenance of the schemes. In particular, we investigate the extent to which socio-demographic differences, institutional variables, and community and group characteristics influence household participation in maintenance of the irrigation systems. It is believed that, once the decision to participate is taken on the basis of certain motivations, these motives together with environmental factors determine household participation behavior. Behavioral decisions carried out by the households concern the amount of time allocated to maintenance activities.

6.3.1 Theoretical model of household labor allocation for maintenance

Drawing on the economic theory of agricultural households (Singh et al, 1986) we develop a model for household labor allocation for maintenance of the irrigation schemes by extending models from existing empirical studies on agricultural household labor allocation (Woldehanna, 2000; Abdulai and Delgado, 1999; Elhorst, 1994; Jacoby, 1993) to incorporate irrigation maintenance as one of the main activities that households in the irrigation communities in northern Ghana perform during the dry season. The attempt is to, at least, partially integrate models on household labor allocation with determinants of effective common-pool resource management.

We assume that during the dry seasons households in the study area allocate labor for three main activities; own farm production (or vegetable gardening) (L^f), maintenance of community irrigation schemes (g_i), and off-farm work (L^o). The standard assumption that the utility function is quasi-convex, twice differentiable and non-decreasing in consumption and leisure apply. Household i has a utility

$$U_i = U(C_{Qi}, C_{Oi}, L_i^l ; A_i)$$ (6.20)

which is derived from the household consumption of farm produce (C_{Qi}), other goods (C_{Oi}), and leisure (L_i^l). We assume that a vector of individual and household characteristics (A) influences preferences. The objective of the household is to maximize its utility. However, the representative household faces some constraints. One is production constraint. The household's technology for the production of dry-season vegetables (Q_i) can be presented as

$$Q_i = Q(L_i^f, \Theta_i, s_i M; D) \tag{6.21}$$

where L_i^f represents household labor; Θ_i a vector of purchased inputs (e.g., fertilizer); s_i is household's share of plots in the irrigable area; D is a vector of group and community characteristics. M is the quality of maintenance of the irrigation scheme (a proxy for quantity of water available in the irrigation scheme).[39] M is assumed to be given by

$$M = f(G, I) \text{ and }$$

$$G = \sum_i^n g_i. \tag{6.22}$$

where G is the sum of labor contributions g_i by all households, also labor contributions by all households other than i can be G_{-i}; I represents the existence of institutions for the management of irrigation systems. The household faces a time constraint,

$$T_i - L_i^l = L_i^f + L_i^o + g_i; \tag{6.23}$$

as well as a full income constraint,

$$Y_i = P_q(Q_i - C_{Qi}) + P_o L_i^o + v - P_c C_{Oi} - P_\theta \Theta \geq 0. \tag{6.24}$$

where T is household total time endowment; Y defines household's full income. Thus, the representative household earns income from farming, off-farm work and remittances (v). P_q is a vector of output prices; P_c is a vector of prices of other goods the household consumes.; P_θ is vector of prices of purchased inputs, and P_o is the market wage for off-farm labor. Households make expenditures on agricultural inputs (Θ), farm products (C_Q) and other goods (C_O). A positive (negative) sign for ($Q_i - C_{Qi}$) implies that the household is a net seller (buyer) of farm produce. The following non-negativity assumptions also hold;

$$C_{Qi}, C_{Oi}, Q_i, \Theta_i, M, G, L_i^l, L_i^f, L_i^o, g_i, G_{-i}, \quad 1 > s_i > 0. \tag{6.25}$$

---

[39] The quantity of water one gets is proportional to the size of his/her plot. So that, the larger the share of a household's plot in the total irrigation area the larger the share of irrigation water

Considering the sizes of plots farmers cultivate at the irrigation sites (mean plot size = 0.143ha), we assume that households hire out but do not hire in labor. Though this assumption is strong, it is reasonable in the context of our sample area because poverty is widespread and farmers are often cash constrained and rarely able to hire labor.

Thus, the solving the first order conditions of the maximization problem outlined above (equations 6.20 – 6.25) yields structural functions of the decision variables, including household maintenance labor supply, in terms of the exogenous variables. Thus, the reduced form notational maintenance labor function resulting from the optimization problem can be expressed as:

$$g = f(P_q, P_o, P_x, P_c, s_i, I, G_{-i}, A_i, D, T_i, v) \qquad (6.26)$$

where the shift factors are as explained above. Input and output prices were observed, but it has been noted that in cross-sectional analysis prices reflect in part regional variations that correlate with households' behavior but are not determinants of their behavior (Dutilly-Diane, Sadoulet and de Janvry, 2003). Prices are themselves endogenous. In particular, output prices can be very low in high production areas and high in low production areas where most households are buyers, in the same way as input prices can be very high in the rural villages. To avoid this endogeneity problem we follow Duttily-Diane et al., to use measures of transaction costs represented by community differences and access to markets, as determinants of behavior instead of observed prices. Other shifters relevant to this model include household characteristics, user group characteristics, institutional variables as well as other external factors. The hypothesized directions of effect are described in section 6.3.3. We attempt to explain the observed variation in maintenance labor contribution across household by testing econometric models specified in the following section.

6.3.2 Econometric model of maintenance participation

In the context of our empirical framework we expand the exogenous variables of the reduced form equation (6.26) to include considerations from the existing literature and country specific variables that address collective action issues raised in this study. Thus, we define a vector of explanatory variables, z, that influence household decision of how much time to allocate to maintenance activities to include: household characteristics, resource attributes, community characteristic (including access to market), profitability, opportunity cost of time for maintenance (i.e., the forgone own-farm labor and off-farm labor income), characteristics of the resource system, institutional variables (rules, norms, etc), as well as user group characteristics (including labor contribution by all other households other than i). More generally, we can write the maintenance labor equation as;

$$g_i = z_i'\beta + u_i \,, \qquad u_i \sim N\left[0, \sigma_{ui}^2\right] \qquad (6.27)$$

where z is as defined above; u_i is the disturbance term.

Maintenance labor allocation behavior varied across households because of heterogeneous preferences. Indeed, the sample contains a significant proportion of observations with reported zero hours of maintenance work, implying that some households fee ride. Thus, the structure for the data generating process suggests that household labor for maintenance is ruled by a latent index g*, that can be expressed as;

$$g_i^* = z_i'\beta + u_i, \quad u_i \sim N(0, \sigma_{u_i}^2) \qquad (6.28)$$

observed only for households who participate in maintenance activities at the irrigation schemes.

Under such conditions, the conventional regression methods fail to account for the qualitative differences between zero observations and continuous observations (Greene, 2003). Restricting our analysis to equation (6.27) will imply involving only those who make a positive labor contribution. This yields biased and inconsistent parameter estimates since the process generating the observed time allocation of farmers is not taken into account (Maddala, 1983). A way out of this is to estimate not only the labor supply function but also other behavioral functions which indicate whether or not a household provides labor for maintenance. It may be necessary to model separately the household decision to participate in maintenance activities at the damsites and the level of participation given the decision to participate.

Indeed, it has been argued that in collective action scenarios, individuals or households must sequentially make at least two decisions (Murdock et al., 2003). First, they must decide whether or not to participate in the collective maintenance activities. Following this participation decision, the individuals must then determine how much to subscribe to the collective good. However, most studies have focused just on the participation choice as though the participation and level of participation decisions can be treated as one (see, Kingma; 1989, Murdoch, 1997; Okamura, 1991). The motivations behind the dynamics in the choices people make are often ignored. We follow this premise to assume that households decide whether or not to participate in maintenance activities and, if the decision is to do so, then how much labor to supply.

We assumed that at the first stage, household i compares the expected utility from participating in the maintenance activities to the utility of not participating. And a household's participation will depend on its utility difference between participation (U_{pi}) and non-participation (U_N), subscript p denotes participation in maintenance over the period under study, and N non-participation. Since by our assumption it is the expected gains in utility from joining that shapes household participation decision, household i participates if its utility difference (h) is positive, i.e., $h_i = U_{pi} - U_{Ni} > 0$. Thus, decomposing household utility into observed (nonrandom) component $(V_{ki}, k = p, N)$ and unobserved (random) component (ε_k), the probability of participation, $\Pr ob_i$ equals

$$\Pr ob_i = \Pr ob[household\ participates] = \Pr ob[h_i \geq 0] \tag{6.29}$$
$$= \Pr ob[V_{pi} - V_{Ni} \geq \varepsilon_{Ni} - \varepsilon_{pi}].$$

Assuming the utility difference is linear in parameters, we specify the utility difference as $h_i = x_i'\gamma + \eta_i$, where $\mathbf{x}$ is a vector of variables that affect the household decision, η is the difference between ε_N and ε_p. That is, the probability that household i provides labor for maintenance

$$\Pr ob_i[participate] = \Pr ob[x_i'\gamma + \eta_i > 0], \qquad \eta_i \sim N(0,\ 1).$$
$$= \Pr ob[\eta_i > -x_i'\gamma] \tag{6.30}$$

The utility difference is, however, unobservable. The unobserved utility difference can be presented as

$$h_i^* = x_i'\gamma + \eta_i \tag{6.31}$$

Therefore, our observation is

$$h_i = \begin{cases} 1 \ if \ h_i^* > 0 \\ 0 \ if \ h_i^* \leq 0 \end{cases} \tag{6.32}$$

and

$$g_i = \begin{cases} g_i^*, if \ h = 1 \\ n.a, otherwise \end{cases} \tag{6.33}$$

Equations (6.28) and (6.31) can be analyzed in a two-stage procedure to identify the determinants of household participation and the level of participation in communal maintenance of irrigation schemes in northern Ghana.

The double-hurdle model, first introduced by Cragg (1971) or Heckman's (1979) sample selection model (also referred to as first-hurdle dominance model in the terminology of Jones (1989)) have been useful in modeling two-stage decision processes (see, Yen, 2003; Joesch and Hiedemann, 2002; Florkowski, et al, 2000; López-Nicolás, 1998; Blaylock and Blisard, 1992; Blundell and Meghir, 1987).

The double-hurdle model features two separate stochastic processes for participation in maintenance activities and level of participation; it allows for the examination of determinants of both participation and level of participation decisions. Thus, to observe a positive outcome two "hurdles" have to be overcome; participation in maintenance (where households develop the desire to participate in maintenance), and to actually contribute labor. Households that pass the participation hurdle ($\eta > -x'\gamma$ in (6.30)), in principle, consider participation as an acceptable option but may not be willing to provide labor for maintenance, for one reason or another. Also those who pass the level of participation hurdle may have no time constraint but may participate in the maintenance activities infrequently and a pattern of zero records can be observed. In the double hurdle model, zeros are attributed to either non-participants or participants who decided not to allocate time for maintenance at a particular point in time, allowing for the possibility of a corner solution. The double-hurdle model assumes correlation between the error terms of participation and the level of participation equations. The model however collapses to what is termed independent double hurdle model when the correlation between the participation and level of participation equations is restricted to zero (Joesch and Hiedemann, 2002), implying that the unobserved characteristics that influence household decisions to participate in maintenance are uncorrelated with the unobserved characteristics that determine how much time households spend on providing maintenance.

The independent double-hurdle model also becomes equivalent to the Tobit model if the probability of passing through the participation hurdle is assumed to be one, which implies that all household are potential participants. In both the double-hurdle and Tobit models, current non-participants are included in the estimation of the level of participation equation. The Tobit model only allows one type of zero observation, namely 'corner solution'. This is based on the implicit assumption that zeros arise because some households will not contribute as a matter of principle. However, since it is against the bylaws of the user groups for one not to participate in maintenance activities it is unlikely that all zeros in the sample represent corner solution. Statistically, the Tobit parameterization also imposes some restrictions on the data-generation process as the same set of variables and parameters determine both the discrete probability of participation and the level of positive participation. Consequently, the probability of a non-zero outcome is tied to the conditional density of the positive observations and this is an undesirable property (Yen and Jensen, 1995; Yen, 1994; Blaylock and Blisard, 1993). The inadequacy of the standard Tobit models in describing situations where different factors determine whether or not households participate makes the Tobit model inappropriate for our analysis.

The first hurdle or Heckman selection model on the other hand, is structured to ensure that all zeros represent non-participation. Unlike the double hurdle or Tobit models, corner solution is not possible in the Heckman selection model. The model is characterized by a latent binary participation equation and a latent level of participation equation. That is, it includes all households in the participation equation but restricts the level of participation equation to households who provide labor (Ward and Moon, 1995). Thus, the Heckman sample selection model addresses the shortcomings of the Tobit model by modifying the likelihood function (as will be shown shortly) to allow a set of variables and coefficients to determine the probability of participation while another set determines the level of participation given that the household participates. The variables in the two models can overlap to a point or may be completely different. The Heckman selection model, like the double hurdle model, assumes correlation between the error terms in the participation and level of participation equations.

Given that we denote zeros in our data as nonparticipation,[40] we use the Heckit[41] to explore the maintenance labor allocation behavior of households at the irrigation communities. It is commonly assumed that η_i and u_i have a bivariate normal distribution:

$$\begin{bmatrix} \eta_i \\ u_i \end{bmatrix} \sim \left(\begin{bmatrix} 0 \\ 0 \end{bmatrix}, \begin{bmatrix} 1 & \rho\sigma_\varepsilon \\ \rho\sigma_\varepsilon & \sigma^2 \end{bmatrix} \right) \tag{6.34}$$

The variance of η_i is set to 1 as it cannot be observed due to the nature of the model (Greene, 1998). Given this assumption, the likelihood function of the model can be presented as:

$$
\begin{aligned}
L &= \prod_{h=0} 1 - \Pr(h_i) \cdot \prod_{h=1} \Pr[\Pr(h_i)\Pr(g_i \mid h_i)] \\
&= \prod_{h=0} [1 - \Pr(\eta_i > -x_i'\gamma)] \cdot \prod_{h=1} [\Pr(\eta_i > -x_i'\gamma)\Phi(g_i \mid \eta_i > -x_i'\gamma)] \\
&= \prod_{h=0} [1 - \Phi(x_i'\gamma)] \cdot \prod_{h=1} \frac{1}{\sigma}\phi\left[\frac{g_i - z_i'\beta}{\sigma}\right]\Phi\left[\frac{x_i'\gamma + \rho(g_i - z_i'\beta)/\sigma}{\sqrt{1-\rho^2}}\right]
\end{aligned}
\tag{6.35}
$$

Taking the log of (6.35)

$$\ln L = \sum_0 \ln(1 - \Phi(x_i'\gamma)) + \sum_1 \ln\left[\frac{\phi(g_i - z_i'\beta)}{\sigma}\right] + \sum_1 \ln\Phi\left[\frac{x_i'\gamma + \rho(g_i - z_i'\beta)/\sigma}{\sqrt{1-\rho^2}}\right] \tag{6.36}$$

where subscript 0 indicates summation over observations with zero hours of maintenance, and subscript 1 indicates summation over observations with positive observed hours of maintenance; $\phi(\cdot)$ and $\Phi(\cdot)$ are the probability density

---

[40] Labor is the only tangible contribution households make towards the maintenance of the schemes studied and it is against the by-laws of the associations for one to fail to participate in a maintenance schedule.

[41] Heckit is an occasional name for generalized Tobit. The Heckit approach allows a different set of explanatory variables to predict the binary choice from those which predict the continuous choice. (The data environment is one in which the continuous choice is measured only when the binary choice is nonzero -- e.g., if we have data on people, whether they bought a car, and how expensive it was, we can estimate a statistical model of how expensive a car other people would buy, but only on the basis of the ones who did buy a car in the data sample.) A regular non-generalized Tobit constrains the two sets of variables to be the same, and the signs of their effects to be the same in the two estimated equations. The 'Heck' in Heckit is for James Heckman. (http://economic.about.com/library/glossary/bldef-heckit.htm)

and cumulative distribution functions of the standard normal distribution, respectively, and $\rho = \text{Corr}(\eta, u)$. However, $\rho = 0$ implies that the two decisions are made sequentially (i.e., one decision to participate is made first and that affect the level of participation decision). This is the dominance model (Florkowski et al. 2000, Jones and Labeaga, 2000). Thus, under the assumption of independence ($\rho = 0$) and there is dominance (zeros are only associated with nonparticipation, not the standard corner solution) the model decomposes into a Probit for participation and standard OLS for level of participation for the subsample of participants, which can be simplified as

$$\ln L = \underbrace{\sum_{h=0} \ln(1 - \Phi(x_i'\gamma)) + \sum_{h=1} \ln \Phi(x_i'\beta) + \sum_{h=1} \ln\left[\frac{\phi(g_i - z_i'\beta)}{\sigma}\right]}_{\substack{\text{Probit} \\ \text{If } \rho=0}} \underbrace{\qquad\qquad}_{OLS} \quad (6.37)$$

Heckman (1979) proposed to estimate the likelihood function by way of a two-stage method as the maximization of the likelihood function took a lot of computing time until recently (see also Puhani, 2000). The expected value of the variable g is the conditional expectation of g* conditioned on its being observed (h=1)

$$E(g_i^* \mid h = 1) = z_i'\beta + E(u_i \mid \eta_i > -x_i'\gamma) \quad (6.38)$$

with

$$E(u_i \mid \eta_i > -x_i'\gamma) = \rho\sigma_\varepsilon \frac{\phi(x_i'\gamma)}{\Phi(x_i'\gamma)}$$

Hence we can rewrite the conditional expectation of g* as

$$\begin{aligned} E(g_i^* \mid h = 1) &= z_i'\beta + E(\eta_i > -x_i'\gamma) \\ &= z_i'\beta + \rho\sigma_\varepsilon \lambda_i \\ &= z_i'\beta + \theta\lambda_i \end{aligned} \quad (6.39)$$

where $\lambda_i = \phi(x_i'\gamma)/\Phi(x_i'\gamma)$ is the inverse Mill's ratio, also referred to as the hazard ratio; $\phi(\cdot)$ and $\Phi(\cdot)$ are as defined, $\theta = \rho\sigma_\varepsilon$, and $\rho = corr(\eta, u)$. Thus, Heckman's (1979) two- stage method is to estimate the inverse Mill's ratio λ by way of a Probit model (6.30) and then to estimate the equation

$$g = Z_i'\beta + \theta\hat{\lambda}_i + \xi_i \quad (6.40)$$

in the second stage, where ξ is an error term summing the influence of all the omitted variables and assumed to be normally distributed with mean zero and constant variance.

As long as η_i has a normal distribution and ξ_i is independent of λ, the Heckman estimator is consistent. It is, however, inefficient if ξ_i is heteroscedastic (Puhani, 2000). Davidson and MacKinnon (1993) suggest the test for selectivity bias by way of conducting t-statistic test on the coefficient of λ, (i.e., since $\sigma_\varepsilon \neq 0$ the hypothesis of $\rho = 0$ can be tested). If the coefficient is statistically equal to zero, then selectivity is not a problem and the sub-sample estimated by OLS regression is expected to yield consistent estimates.

6.3.3 Variables and hypotheses

The dependent variable measuring the extent of household participation in the maintenance of the irrigation system is defined in terms of labor hours for maintenance. The number of days one participated may not be the exact measure since the number of hours of maintenance work per day could be varying across households. In the following paragraphs we provide some short comments on the hypothesized direction of effect of the variables used in the econometric analysis of household incentive to participate in the collective maintenance of the schemes. Table 6.9 presents definitions and summary statistics of the variables used in the analysis.

It is assumed that the household decision to partake in maintenance activities is affected by household characteristics, community characteristics, the opportunity cost of labor, characteristics of the irrigation system, as well as economic and institutional factors. These factors are often identified as being useful for understanding cooperation in collective action (Dutilly-Diane et al, 2003). Factors that are specific to the household include household composition (comprising members less than fifteen years of age and members with ages fifteen and above), and characteristics (age, education of the head, etc). Other factors include profitability of household garden, resource endowments (plot size or household share of plots at the scheme, off-farm income, size of household arable land outside the scheme), institutional variables (rules and norms), and community characteristics (including location, market, economic opportunities) among other variables also described below.

Due to previous experiences of broken down irrigation schemes, older people are expected to have a better understanding of the consequences of non-participation in maintenance. The older people who experienced the consequences of broken down schemes, which resulted mostly from lack of maintenance, are expected to be motivated participants. However, a negative effect is expected when households act as in the past when government was responsible for maintenance and resist the new paradigm where farmers are asked to take up maintenance responsibilities. The direction of the effect of household head's age on the participation decision is thus indeterminate, *a priori*. The effect of age may be nonlinear, so we include the square of the age of the household head to examine how household contribution to maintenance will change over the lifecycle of the head.

A priori, the direction of the effect of education on participation in collective maintenance is not very clear. Education improves household access to information and promotes adoption of technology to make farming profitable.

Coupled with the external recognition, education enhances participation in collective action (Meinzen-Dick et al., 2002). Education may, however, increase opportunities outside agriculture and for that matter lower participation in the collective action. On the other hand, alternative incomes earned through increased opportunities may help the farm households pay for purchased inputs, settle irrigation levies and for that matter enhance their participation in collective action. The foregoing reasoning could also be true for households with alternative income sources.

Given the difficulty of collecting reliable data about the cash income households earn from off-farm activities, we assumed binary information regarding whether or not the household engaged in non-farm economic activities. The expected direction of effect of off-farm income is ambiguous as stated above. Households with more income opportunities may be facing higher opportunity costs of labor, as a result of which they may be more reluctant to allocate more time to maintenance.

Ownership of arable land outside the irrigation scheme is posited to have a negative effect on participation. Households with larger land areas may be able to produce more from the main (wet) season to meet household food and income needs. Such households may be less interested in the scheme.

Profitability of the household's activities at the irrigation scheme will be a major motivator for participation in the maintenance of the scheme. Profitability is considered to be an endogenous variable. It may depend on plot size or household share of land area at the scheme, household and community specific variables as well as cultivation of high value crops, among other variables.

Land holding being the main unit of economic activity at the irrigation schemes affects households' participation in collective maintenance in a decisive way. Households with larger plots are expected to have greater incentives to make investments (contribute to quality maintenance for the sustainability of the schemes) in anticipation of greater benefits. Thus, it is assumed that the larger the size of a household plot, the greater the expected benefit and the more the incentive to make positive contributions to maintenance. Total landholding is the size of plots (in hectares) operated by the household at the irrigation scheme for gardening during the 2002/2003 dry season. Also, scheme size is expected to have a negative correlation with the level of participation. Monitoring will be easier in smaller schemes than in larger schemes.

Household perception about fairness in the distribution of plots can affect its motivation to participate in the collective maintenance of the schemes. Plot Gini is a measure of inequality in plot allocation across the irrigation schemes whose direction of effect is not clear a priori, as there are counteracting effects. On the one hand, larger plot owners may act as leaders and engage in more maintenance themselves. On the other hand, small plot owners may free-ride. Also, the household's perception about the quality of its plots relative to other plots is expected to have a positive effect on their contributions to maintenance.

Table 6.9: Descriptive Statistics

Variable	Mean	Std Dev.
Age of Household head (years)	42.50	15.81
Age of Household head squared	2056	1528.19
Household size (number)	9.40	4.92
Household members < 15 years old	3.90	2.65
Household members ≥ 15 years old	5.50	3.38
Literacy (literate = 1 when attained more that six years of education; 0 otherwise	0.27	0.45
Off-farm income of head (earned more than 100 thousand cedis in the dry season; 0 otherwise)*	0.23	0.42
Plot sizes	0.14	0.18
Participation rule (1= there is a rule that enjoins people to participate in communal activities; 0 otherwise)	0.67	0.47
Forfeit of plot for failure to participate (1=heaviest penalty for nonparticipation is forfeiture of plots plot; 0 otherwise)	0.08	0.27
Fine(1=penalty for non participation is fine; 0 otherwise)	0.51	0.50
Size of land outside scheme (Ha)	2.00	1.88
Labor intensive (1=method for dam rehabilitation was labor intensive; 0=capital intensive (contractor))	0.72	0.43
Quality of service (1=satisfied with delivery schedules; 0 otherwise	0.66	0.47
Plot gini (gini coefficient)	0.50	0.10
Accountable leadership (1=leaders render regular accounts; 0 otherwise)	0.60	0.50
Market access (1=periodic market in community; 0 otherwise	0.63	0.48
Distance to market (km)	1.47	2.49
Size of scheme (ha)	12.48	8.19
Bad opportunities (1=bad opportunity of earning alternative income, 0 otherwise)	0.80	0.40
Water shortages (1=experienced water shortage in the past season; 0 otherwise)	0.43	0.49
Quality of the scheme (1=satisfied with quality scheme rehabilitation work; 0 otherwise)	0.32	0.11
Age of WUA (years)	3.68	3.42
Shadow wage (shadow wage of maintenance labor(cedis/hr)	1884.77	3101.61
Social interaction (households per ha)	22.97	21.06
Plot is located at the head (=1; 0 otherwise)	0.25	0.44
Plot is located in the middle (=1; 0 otherwise)	0.48	0.50
Plot is located at the tail (=1; 0 otherwise)	0.27	0.44
Community is urban (=1; 0 otherwise)	0.021	0.14
Community is semi-urban (=1; 0 otherwise)	0.06	0.24
Community is a village (=1; 0 otherwise)	0.82	0.38
Community is a peri-urban (=1; 0 otherwise)	0.10	0.29

* Cedi is the Ghanaian currency. (US$1=8600 cedis at the time of survey)

Enforcement of rules (or bylaws) of user groups can enhance participation. Weak or poor rule enforcement provides less incentive for households to contribute to the collective action required for the maintenance of the scheme, while a credible threat of punishment if rules are not complied with offers incentives to participate in maintenance activities. We expect an increased level of household participation in maintenance where there are rules governing participation of individuals in communal activities. Greater participation is expected of households in user associations which enforce rules that ensure forfeiture of plots for failure to participate in maintenance activities.

Accountability is a measure of transparent leadership. We expect households in user groups whose leaders are perceived to be accountable to be positive in their response to calls for collective maintenance activities.

The technique used in the rehabilitation program can affect cooperative behavior of households. The mode of rehabilitation that afforded members the opportunity to directly participate in the rehabilitation process acquainted the farmers with operation and maintenance techniques. Households whose members directly participated (provided labor) in the rehabilitation process are more likely to actively participate in maintenance.

The quality of the construction or rehabilitation work done by contractors affects farmers' attitude towards maintenance. If dissatisfied with quality of rehabilitation, communities tend to attribute deteriorations to poor quality of work done by the contractor. They are reluctant to participate in maintenance. They expect the contractor to be brought back to put things right. We expect a positive relationship between the quality of rehabilitation work and household maintenance participation decisions.

Water shortage is expected to have a positive effect on household participation decisions and maintenance efforts. Past experiences and consequences of the collapse of previous irrigation systems as a result of lack of maintenance make water shortages a perfect reminder of the past and a signal to take action. Household perceptions about the quality of service (satisfaction with water delivery) are also expected to have a positive effect on the contribution to maintenance effort.

Higher social interaction among irrigators made possible by higher population density is important for organizational enforcement of collective action norms, by enhancing peer monitoring. Social interaction is assumed to be intense if more people are concentrated in a small area (Fujita, Hayami and Kikuchi, 1999). The number of households per hectare of the irrigable land is used as a proxy for the degree of interactions among households. At schemes where social interaction is high, peer monitoring is high and household participation in maintenance is expected to be high.

Market access is expected to have a negative effect on household participation in collective maintenance of the irrigation schemes. Increased market access may create exit options and increase the opportunity cost of time. The important social role that markets play in the lives of the people in the study area (section 2.3.3) would mean that maintenance schedules coinciding with market days would not receive a positive response.

In this study we characterize communities in terms of the difficulty of earning alternative incomes. The presence of alternative income earning opportunities increases the opportunity cost of labor for maintenance and provides exit options in the event of a scheme breakdown. We therefore expect households with fewer opportunities to work off-farm to provide more labor for maintenance.

Wage represents the opportunity cost of labor for maintenance activities at the scheme. A negative relationship, where high wages reduce labor supply for maintenance, will result where there are exit options and the opportunity cost of labor is high. Thus, it is expected that high wages will have a negative effect on household labor contribution to maintenance.

Theoretically, households will allocate labor between farm work and non-farm activities (maintenance and off-farm, in our broad definition) in such a way that the marginal products of labor for these activities are equal, and equal to the wage of the off-farm activity. However, the estimated marginal value products (shadow wage rates), especially of maintenance, may not be equal to off-farm wage or farm wages under labor market imperfection or when the family does not sell labor. Besides, liquidity constraints may results in a situation where farmers engage in off-farm activities in the slack season to finance activities in the farm sector (see Woldenana, 2000). Therefore, market wage cannot be a suitable substitute for shadow wage in the estimation of the household labor for maintenance function. The shadow wage rate (w_g) per hour for household i is derived based on a Cobb-Douglas production using the expression;

$$w_g = \frac{\hat{Q}_i(\cdot)}{g_i} \cdot \hat{\gamma}^g \qquad (6.41)$$

where $\hat{Q}_i(\cdot)$ is the fitted value of output by household i, g_i is hours of labor supplied by household i for maintenance activities, and $\hat{\gamma}^g$ is the estimated coefficient of household contribution (labor) to the public good in the production function. The results of the Cobb-Douglas production function[42] fitted are presented in Table D6.4. Examples of this approach can be found in Jacoby (1993), Skoufias (1994), Abdulai and Regmi (2000), Woldehanna (2000) and Mensah-Bonsu (2003) where the authors substituted marginal product of labor for shadow wage.[43] A drawback of this method is that shadows wage rates can be endogenous to the system and any changes to the underlying determinants can result in different optimal levels of shadow wages. We tried to overcome the simultaneity problem by using predicted values of the shadow wages (Table D6.5) in the analysis.

---

[42] The Cobb-Douglas functional form was used despite the well-known technological restrictions it imposes, because a large proportion of the marginal value products estimated using a more flexible (quadratic) functional form turned out to be negative. Besides, we thought that due to factor market restrictions that farmers in rural communities face perfect factor substitution cannot be expected. We therefore assumed that the Cobb-Douglas framework can be appropriate for our study.

[43] A cursory examination shows the estimated marginal productivities to be higher than the observed wages for off-farm labor. This appears to suggest that the use of observed market wages in the analysis could result in underestimation of the shadow wages.

When water is not equitably distributed, spatial factors affect participation behavior of water users. Thus, the location of the farm in the scheme can be thought to have an influence on the incentive to participate in maintenance, as benefits to members located at the tail end will be considerably different from benefits received by those located towards the head. This can result in rational disinterest of tail enders (downstream irrigators) to participate in maintenance. Locality dummies are also included to examine how type of community (urban, semi-urban, and rural) or its proximity to urban center (peri-urban) affects household behavior towards maintenance.

6.3.4 Results and discussions

This section presents the results of the econometric model outlined earlier (section 6.3.2). We treat those who provide labor and non labor providers in a systematic fashion. At the maintenance stage where households supply different amounts of labor hours based in part on expected utilities to be derived, which in turn drive their response to obligations as members of the WUA. The Heckman two-stage sample selection model was estimated by maximum likelihood methods. The correlation coefficient (Rho) was found to not be statistically different from zero (Table D6.6). The fact that Rho is not significant (statistically different from zero) means that selectivity is not an issue and that we can analyze the two situations separately on this data. For the purposes of clarity we present estimates from the probit and OLS regressions separately. We first present the results of household participation decisions based on a Probit regression. We estimate robust standard errors in an attempt to address the problem of any heteroscedasticity induced or present in the cross-sectional data.

6.3.4.1   Participation decision

As a first step in understanding how households in our dataset respond to their maintenance obligations at irrigation schemes, we estimate a Probit regression of household decisions to participate in maintenance activities organized by the water users' associations. The estimated coefficients are presented in Table 6.10. The signs of the coefficients of the regression are important to us, as they indicate the direction of change in the probability of the dependent variable taking a particular value, given the change in the independent variables. The χ^2 value of the model was significant at the 1% level indicating that the independent variables taken together influence households' participation in the maintenance of the irrigation schemes. The McFadden (pseudo) R-square is 0.48.

The square of the age of the household head was included in the analysis to show lifecycle effects on household decisions to participate in maintenance activities. The results indicate that the probability of household participation in maintenance activities increases with the age of the head until it reaches a maximum around the age of 45, and declines thereafter. Literacy status of the household head had a positive and significant effect on the probability of participation. The ability of the household head to read or write increased the probability of household participation in collective maintenance activities as education improves his/her ability to appreciate the importance of collective action. When household off-farm income increases, the probability of household participation also increases. This effect is unexpected. Probability of participation is also high for schemes that were rehabilitated by labor intensive technique, especially for households whose members were involved in the rehabilitation program. Water shortages increase the probability of household participation in maintenance. Perhaps water shortages become a reminder of past experience of broken down schemes that resulted from poor maintenance or lack of it.

Table 6.10: Determinants of household participation in collective maintenance

Variable	Coef.	Robust S.E	P>| z		Marginal Effects˙ Coef.	S.E	P>| z	
Constant	-2.54978	0.58145	0.000			
Age	0.03260	0.01925	0.090	0.01230	0.00728	0.091
Age Squared	-0.00036	0.00019	0.065	-0.00014	0.00007	0.066
Hsehold mem < 15 years	-0.01598	0.02582	0.536	-0.00603	0.00975	0.537
Hsehold mem ≥15 years	0.00020	0.01554	0.990	0.00007	0.00586	0.990
Literacy	0.34241	0.13459	0.011	0.12914˙	0.05077	0.011
Off-farm income	0.25442	0.14855	0.087	0.09336˙	0.05258	0.076
Forfeiture of plots	0.29891	0.19744	0.134	0.10594˙	0.06666	0.112
Land outside scheme	-0.00757	0.01026	0.461	-0.00286˙	0.00387	0.460
Contr. labor during rehab.	0.81327	0.12876	0.000	0.31214˙	0.04811	0.000
Satisfied with quality of service	0.29746	0.10549	0.005	0.11217	0.04009	0.005
Plot Gini	-0.29858	0.55991	0.594	-0.11261	0.21122	0.594
Accountable leaders	0.64091	0.17488	0.000	0.24092˙	0.06437	0.000
Market access	-0.03224	0.02473	0.186	-0.01216	0.00919	0.186
Bad opportunity	0.19192	0.19233	0.318	0.07355˙	0.07468	0.325
Water scarcity	0.44479	0.14939	0.003	0.16462˙	0.05332	0.002
Quality of rehab. work	1.11749	0.64547	0.083	0.42146˙	0.24325	0.083
Age of the WUA	0.10056	0.03303	0.002	0.03793	0.01234	0.002
Wage per hour	-0.00007	0.00014	0.649	-0.00003	0.00006	0.649
Social interaction	0.00416	0.00314	0.185	0.00157	0.00118	0.185
Others participate	0.01973	0.09188	0.830	0.00744	0.03465	0.830
Profitability	1.77447	4.11084	0.666	0.66924	1.55063	0.666
Community is urban	0.01890	0.31431	0.952	0.00711˙	0.11783	0.952
Community is peri-urban	0.10390	0.21435	0.628	0.03854˙	0.07828	0.622
Community is rural	0.10157	0.26259	0.699	0.03773˙	0.09592	0.694

No. of observations =821; Wald Chi-sq. (24) = 349.97; Prob. > Chi-sq. = 0.0000; Pseudo R = 0.4853
(˙) dy/dx is for discrete change of dummy variable from 0 to 1.

Perceptions about the quality of the rehabilitation work have a positive and significant effect on the probability of participation in maintenance activities organized by the WUA. Households in communities that were not satisfied with the quality of work done by the contractors who rehabilitated the schemes are apathetic towards maintenance. Often deteriorations are attributed to poor quality of work, and households would not want to repair the systems until the contractor is brought back to fix it.

The results show that the probability of household participation in maintenance is higher in WUAs with more transparent and accountable leaders. Accountability promotes trust and goodwill that increases the success of collective action. The result is consistent with previous observations about the importance of trust for successful self-organization (Ostrom, 1994). An observation to this extent was made, during our survey at Kadi (one of the survey communities), where lack of transparency, and the resulting leadership crisis, has led to the near collapse of the scheme. Probability of participation also increases in older WUAs as a result of accumulated experience in community mobilization and, more importantly, increased familiarity and social capital gained by group members.

The explanatory variables that are not statistically significant in explaining the probability of participation in maintenance are equally important. The results suggest that households with a smaller adult population (more of its members being children and under the age of 15) show a decreased probability of participating in maintenance activities. The converse is true for households with many dependents. This result is not statistically significant, though. Inequality in plot allocation seems to be a drawback to participation in collective maintenance. Though not statistically significant, the direction of the effect tends to corroborate earlier findings that have supported this view (Agrawal, 2001; Dayton-Johnson, 2000a; Baland and Platteau, 1999, 1996). The expected negative effects of household ownership of arable land outside the scheme, distance to markets, as well as the negative effect of wage rates on the probability of participation are indicative but not statistically supported by our data set. The institutional variables (participation rule, and plot forfeiture norm) and also the variable representing lack of alternative opportunities bear the expected signs but do not have a statistically significant effect on households' decision to participate in maintenance. None of the locality variables was statistically significant suggesting that the location of one's community has no influence on maintenance participation decisions.

6.3.4.2   Effort Contribution

We now turn to the second stage household decision model to attempt to explain effort contribution (how much labor time a household contributes), conditional on participation in maintenance activities. The predicted coefficients of the second stage regression are presented in Table 6.11 below. The F-statistic is significant at the 99% level, supporting the joint hypothesis that all the non-intercept coefficients are significant in explaining the observed variations in the level of labor hours contributed by households.

The independent variables examined include many of the variables included in the first stage equation. The household specific characteristic effects of the Probit model carry over. Age of household head and other household demographic factors were all significant in explaining the variations in the level of participation across households. Pronounced and intuitive lifecycle effects emerge as households contribute more and then less labor up through about the age of 51, indicating that the amount of labor hours contributed by households increases with the age of the head, reaches a maximum level at the age of 51 and declines thereafter as the head increase in age.

Maintenance labor was positively and significantly affected by household size. In particular, households with less labor constraints (many adult members) offer more labor for maintenance work. All things being equal, a unit change in the size of the adult population could change the household's labor contribution by 0.2. The results also suggest that households with many dependents (fewer adults) provide less labor. For households who participate in maintenance, profitability is one of the main determinants of the number of hours they offer for maintenance work at the scheme. Profitability positively influences the number of hours of maintenance work households perform. This implies that households that obtain higher returns from gardening offer more labor for the maintenance of the scheme.

Contrary to the *a priori* expectation, off-farm income of the head of households who participates in maintenance, positively and significantly affects the number of hours the households offer for maintenance. This is also consistent with the result of the participation equation which indicates that households with off-farm income contribute more labor for the maintenance of the scheme. The reason may be that the seasonality of irrigation activities may allow households to participate in off-farm activity to gain alternative sources of cash to finance dry season gardening at the irrigation schemes (Skoufias, 1993). Besides, these households may be able to hire labor to meet shortfalls their participation in maintenance may cause. Indeed, in a region such as the study area where access to farm credit is very poor, off-farm income increases the ability of farming households to pay for purchased inputs. This suggests a complementary relationship between off-farm income and participation, especially when the credit market is imperfect.

Satisfaction with water delivery schedules has a positive and highly significant effect on the level of participation. Members who are dissatisfied with water delivery schedules are not well motivated to contribute to maintenance efforts. Households who do not get enough water when it is most needed resort to digging of shallow wells and thus become less dependent on the irrigation facility.

Rules and regulations condition household participation in collective action. Our results show that enforcement of rules and norms limits free riding and promotes collective action. Norms such as forfeiture of plots for failure to participate in maintenance tend to deter free riding. The size of household land outside the scheme has no significant effect on participation. Land ownership could provide an indication of the level of food insecurity and deprivation suffered by households, though. If the sign on the variable is anything to go by, then it could indicate that households with larger plots outside the schemes may have enough resources to feed their families and therefore remain passive in regard to the sustainability of the irrigation schemes. Large upland owning households may be able to produce more during the main season to meet their food and income needs.

Table 6.11: Determinants of level of household participation in collective maintenance

Variable	Coefficient	Robust S. E	P>| t	
Constant	1.12752	0.37256	0.003
Age	0.02772	0.01047	0.011
Age Squared	-0.00027	0.00011	0.057
Household members < 15 years	0.00570	0.01294	0.660
Household members ≥15 years	0.03616	0.01004	0.000
Profitability	3.54177	1.85849	0.057
Off-farm income of household head	0.16926	0.07357	0.022
Forfeiture of plots	0.17191	0.10338	0.097
Land outside scheme	-0.00273	0.00560	0.626
Cont. labor during rehabilitation	0.15559	0.07407	0.036
Satisfied with quality of service	0.10736	0.06072	0.078
Plot Gini	0.23514	0.22621	0.299
Accountable leaders	0.21158	0.11543	0.068
Market access	-0.15103	0.07982	0.059
Water scarcity	0.14452	0.07538	0.056
Size of scheme	0.01462	0.00449	0.001
Age of the WUA	0.02012	0.01713	0.241
Shadow Wage per hour*	-0.00016	0.00002	0.000
Social interaction	0.00628	0.00156	0.000
Plot is located at the head end	-0.01322	0.07623	0.886
Plot is located at the tail end	-0.04053	0.07572	0.593
Community is urban	0.04723	0.22827	0.836
Community is peri-urban	-0.13135	0.10815	0.225
Community is semi-urban	0.08413	0.13976	0.584

No. of observations = 445; $F_{(24, 420)}$ = 7.62; Prob. > F = 0.0000;
R-squared = 0.4478; Root MSE = 0.6518

* Predicted values

Water shortage is another variable that significantly affects households' incentive to offer more labor for maintenance. The variable has a positive and significant effect on maintenance labor allocation. For many of the communities in the study area, the dams remain the major sources of water, underscored by its multiple uses. For that reason, water shortages have a severe impact on households in the communities. Past experiences and consequences of broken down schemes which resulted from poor maintenance, make water shortages perfect signals that encourage households to provide more labor for maintenance. Scheme size also positively and significantly affects the level of household participation in maintenance, contrary to our expectation. Market access has a significant negative effect on household contribution to the maintenance effort. As hypothesized, increased market access raises the opportunity cost of labor. Also, opportunities to earn alternative incomes through market participation would reduce interest in the schemes, while social events associated with markets in the study area draw a large number of people to the market centers, especially during market days.

Our results also indicate that perceived accountability of the leaders of the user group positively influences the level of participation. This is also consistent with the results of the participation decision function. The extent of participation in maintenance activities is high among households who perceive WUA leaders to be accountable. The effect of a shadow wage on the level of participation is consistent with our *a priori* expectation. Increased opportunities for labor and resource exchanges outside the village tend to be less desirable for cooperative behavior, as availability of exit options raises the opportunity cost of labor. Households offer less labor hours for maintenance when wages (the opportunity cost of participation) are high, all things being equal. Social interaction positively affects the level of participation. Consistent with our *a priori* assumption, peer group effect galvanized in higher social interaction has a positive and significant effect on the level of participation. A high level of interactions promotes peer monitoring and reduces the tendency for free-riding. The variable representing the use of labor intensive techniques is positive and significant. Physical participation of household members in the rehabilitation process tends to encourage higher effort contributions to maintenance as a result of the learning effect of the program. Neither of the dummy variables for head-tail location of household plots nor the locality dummies had a discernible effect on the level of participation, suggesting that location of ones plot or community did not matter.

6. 3.5 Concluding remarks

In this section we have focused on analyzing the factors that influence the decision of households to participate in collective management of irrigation schemes in northern Ghana, and the total amount of time allocated to maintenance work, if the decision is to do so. We have presented and estimated models for household participation decisions and the level of participation in collective maintenance of the irrigation schemes in a two-stage framework. Our results shed some light on the reasons why households differ in their contributions to the maintenance of the community irrigation schemes.

As hypothesized, household participation in collective maintenance is affected by household characteristics, community characteristics, institutional variables and some characteristics of the resource. In particular, this study shows that household labor contribution exhibits a U-shaped relationship with respect to the age of the household head. Household Labor endowments appear to matter. Households with more adult members contribute higher amounts of time for maintenance. The finding is also consistent with the assertion that profitability is one of the main motivations for participation in collective action. The labor intensive technique of rehabilitating the schemes, in which beneficiary communities were allowed to contribute labor, increases both the likelihood of participation and the amount of time spent on collective maintenance of the schemes.

The analysis revealed that accountability of WUA leaders is a significant factor conditioning households' willingness and extent of participation in collective management. Participation in collective action is high where leaders' activities are more transparent and leaders are more accountable, periodically declare accounts of their stewardship and seek approval for all financial transactions. The size of the schemes and for that matter plot sizes have a motivative effect on household participation in maintenance activities. Households in larger schemes have relatively larger plots and the expected returns to efforts are higher, *ceteris paribus*. Inequality in plot allocation appears to reduce the likelihood of participation in collective action for the maintenance of the schemes. The results also revealed that access to markets and existence of exit options may reduce the incentive to participate in the collective maintenance of the scheme as they increase the opportunity cost of labor. Water scarcity is positively correlated with participation in collective maintenance as multiple uses of the irrigation schemes become a motivating factor for increased participation in maintenance activities. Households in communities that experience perennial water shortages have greater motivation to maintain the schemes due to the multiple uses of the schemes.

The analysis further revealed that local institutions play a very important role in enhancing household participation in collective activities. Enforcement of bylaws such as forfeiture of right to farm in the scheme for failure to participate in collective maintenance activities, as well as social norms guarded by peer monitoring mechanisms encourage greater participation. Poor or selective enforcement of bylaws can perpetuate free riding. Since institutions help to guide human behavior, clear rules regarding participation are a key to successful community-based resource management.

In sum, the analysis in this section showed that household decisions to participate in collective maintenance, and the amount of time actually allocated to the maintenance of the community irrigation scheme, are independent, and are influenced by household characteristics, institutional factors, as well as community and group characteristics, which converge with the factors determining the success of collective action at the community level. Addressing these issues can greatly improve the outcomes of community management of the irrigation facilities.

7 CONCLUSIONS

7.1 Summary and conclusions

In many countries, institutional weaknesses and performance inefficiencies of public irrigation agencies have led to high costs of development and operation of irrigation schemes. Poor maintenance and lack of effective control over irrigation practices have resulted in the collapse of many irrigation systems. Moreover, irrigation agencies have largely failed to raise sufficient revenues from the collection of water charges to meet operational expenses. Consequently, there has been growing promotion of community-based irrigation management in many developing countries to improve efficiency and reduce cost.

Evidence of successful devolution programs in some parts of the world has motivated a lot of donor support for small-scale irrigation schemes under community management. Leading this initiative in Ghana, the International Fund for Agricultural Development (IFAD) has funded the construction and rehabilitation of communally managed dams in the Upper East Region under the Land Conservation and Smallholder Rehabilitation Project (LACOSREP), with ownership rights and responsibilities transferred to beneficiary communities. The success story of LACOSREP has engendered the replication of this model by other NGO and donor funded programs with irrigation components in northern Ghana.

However, the actual outcomes of devolution programs in various countries have been mixed, and the stated objectives of achieving positive impacts on resource productivity, equity, poverty alleviation and environmental sustainability are often not met (Knox and Meinzen-Dick, 2001). Evidence of success, especially in the smallholder context, remains limited. Indeed, previous experiences with community managed irrigation schemes in northern Ghana have not always been positive. In fact, many schemes severely deteriorated or broke down completely in the past due to insufficient maintenance and lack of catchment area protection. Therefore, an understanding of the outcomes of community management of irrigation schemes and the reasons why communities differ in terms of economic, environmental and distributional outcomes of irrigation management is essential for making improvements in irrigation management at the community level.

Attempts are therefore made, in this study, to investigate the factors that affect the success of community-based management of irrigation schemes in northern Ghana. To do so, the study: (1) assesses household incentives to participate in collective action for the maintenance of irrigations schemes, (2) examines the outcomes of community-based strategies for managing irrigations schemes, and (3) analyzes the conditions for successful community-based irrigation management in northern Ghana. The goal is to improve the understanding of the conditions for long term sustainability of community-based irrigation management strategies, in terms of sustainability in maintenance and sustainability of local institutions. It is hoped that the results of the study will help to bridge the gap between theoretical predictions and empirically observed factors that condition the success of community-based management of natural resources.

The focus of the study was in the Upper East Region (UER) and Upper West Region (UWR) in northern Ghana, which are characterized by low and erratic rainfall regimes, low soil fertility, rapid soil degradation, high population densities and extreme poverty. Agriculture is the main source of livelihood for the majority of the population living in the study area. These regions boast of large numbers of small-scale irrigation schemes, and are in the forefront of community-based irrigation management in Ghana. It is hoped that the research findings provide insights that can guide similar efforts, even in different resource sectors, in other parts of the country with similar agro-ecological and socioeconomic settings.

The study begins with a review of the socio-economic contingencies of the study area which could enhance or hinder the success of community management of common-pool resources, including household organizations and labor exchange arrangements. It was noted that communal labor exchange arrangements that abound in the area lead to development of social networks, norms and trust upon which people tend to cooperate for common welfare. One other important aspect of the community system relevant for the success of collective action is the fact that the patriarchal household arrangements prevailing in the area make social mobilization easy as households play important roles in mobilizing their members for social and collective activities in the communities. The review identifies a great variety of socio-political and traditional institutions that play important roles in natural resource management in the region. Indigenous people's knowledge, practices, values and capabilities form the basis for natural resource management in the local communities. Traditional institutions function to activate and enforce social norms of behavior through established patterns of authority and leadership. Thus, individuals in the local communities are constrained by ethical and moral values that induce cooperative behavior. However, we find that while chieftaincy disputes in the area threaten social cohesion and stability, the declining influence of traditional systems and changing perceptions (including belief systems) about the spirituality of natural resources (due to external pressures e.g., globalization, migration, access to television and the media, introduction of western education and religion, etc.) weaken the ability of the traditional institutions alone to promote sustainable natural resource management. It was suggested that strengthening the capacities of the traditional authorities to develop systems of cooperation and effective systems of monitoring and enforcement of community regulations can be a key to the success of natural resource management at the community level.

An account is also made of the evolution of participatory irrigation management in northern Ghana to provide an understanding of the impact of institutional and organizational efforts on the efficient management of the irrigation schemes. Indeed, most of the prominent interventions in agricultural development in the study area have been in irrigation. Several dams were constructed in the study area to provide water for dry season gardening and livestock watering. In the past, various organizational forms of state management of these schemes had been fraught with inefficiency and total neglect that led to the collapse of several of the schemes, the majority of which are being rehabilitated for communal management. Experiences from past failures have informed current interventions and the general policy of natural resource management transfer in Ghana. The structure of the local unit of organization, the water users' association (WUA), entrusted with the management of the irrigation schemes was analyzed to provide an understanding of the structural framework of the WUA for carrying out its functions. Past experiences of deterioration and collapse of irrigation schemes in the study area make scheme maintenance the most important of the functions of the WUAs, which also include distribution of irrigation plots and water, collection of irrigation fees and resolution of disputes. Through collective action, the WUAs mobilize resources for operation and maintenance of the schemes. However, user groups differ substantially in their performance of these functions, with some systems exhibiting very low levels of maintenance.

The study draws on a field survey conducted in the first half of 2003, covering irrigation schemes under community management in all the districts of the Upper East and Upper West regions. Two rounds of survey were conducted. The first round involved technical evaluation of the state of the irrigation schemes to serve as the principal indicator of the user groups' success in mobilizing resources for the maintenance of the irrigation facilities, using a checklist of irrigation system quality measures. The second round covered 52 communities with functional irrigation schemes, and comprised community and household surveys. Households were stratified into members and non-members of the irrigators association. The non-members included only those households who did not have any member practicing irrigation at the community scheme. In all, 302 non-members and 520 members were interviewed. The sets of questionnaires developed for the survey covered elaborate socioeconomic information about the households and the communities as well as the structure of the WUAs, functions, norms, and rules for distribution of costs and benefits.

It was observed that dry season gardening practiced at the irrigation schemes plays a very important role in the livelihood strategies of most households in the study area. The dams also remain the most important sources of water for livestock in the area, and for other domestic uses in addition to fishing. The multiple uses of the schemes serve as the incentives for members of the communities to participate in the collective maintenance of the systems. However, as common property the schemes are open to the problems of collective action (e.g., free riding). The tendency for some households to renege on their maintenance responsibilities often result in a cycle of insufficient maintenance and declining performance.

On the basis of the community dataset, the study analyzed the outcomes of community management strategies typifying the level of collective action at the community level, and the factors that facilitate these outcomes. We defined the outcomes of collective action to include: rules for distributing costs and benefits, concerted efforts by user groups to comply with rules, and quality of maintenance achieved by the community through collective action.

To ensure that the economic benefits of the irrigation schemes are equally shared by all, the beneficiary communities were called upon to agree to a land redistribution policy that would ensure that the majority of the people, including women and landless, have equal access to irrigation land. However, some landlords continue to resist the land redistribution policy. In the majority of the schemes plots are fragmented to accommodate large numbers of community members. Plot sizes ranged from 0.004 – 1ha. The mean Gini coefficient, measuring inequality in plot distribution was 0.35, and ranged from a maximum of 0.72 to a minimum of 0.03. The extent of inequality, however, varied across communities.

On the other hand, mechanisms for distributing water at the schemes were prescribed on the principles of equal distribution, but in principle it is in proportion to shares of landholdings in the irrigation area. Irrigators with larger plots get proportionally more water. Water distribution arrangements varied across the schemes and took the form of either a continuous flow arrangement or a time rotation mechanism. A dichotomous choice model is used to analyze the determinants of the choice of water distribution rule. Number of villages sharing the use of the same scheme, age of the user group, conflicts and resistance of landlords to land redistribution increase the probability that a user group chooses the rotational system of water distribution. However, there is higher probability of observing continuous flow arrangements in communities that are socio-culturally heterogeneous. A potential explanation of this result is that the extent of organization and coordination required under a rotational mechanism will be difficult to achieve in heterogeneous communities. Continuous flow arrangements are more likely to be adopted where communities have "illegally" extended the schemes, and where water shortages have occurred. Perhaps, this is an attempt to avoid the likely event of the water getting finished before it gets to the turn of others in a rotation system.

There are distributional implications in the allocation of land and water, which can be critical to the success of collective action for the management of the irrigation schemes. When resource allocation is perceived to be unfair it could lead to practices that can derail cooperative efforts and render the institutions for managing the resource dysfunctional. The study shows that land conflicts are more likely to emerge where irrigation is more profitable, group sizes are large and where landlords have resisted the land redistribution policy.

Lack of opportunities for alternative sources of income also increases the probability of conflict occurrence. It was, however, noted that there is a low probability of land conflicts where the number of original landowners who are to give up their lands is large. The moral pressure on landlords to give up their lands for redistribution could be high when others have already done so.

The causes of conflicts over water are inter-related. Water conflicts in this study are highly associated with frequent water shortages, size of the command area and unapproved extensions to the systems. Demand for water in schemes with larger command areas may go beyond the reservoir capacity, and result in water shortage. Also of significant effect on probability of conflict occurrence is the water allocation rule. Disputes over water are more frequent in schemes where the time rotation mechanism for water distribution is employed, due to the tendency for others to illegally open water to their fields when it is not their turn. Improvements in the conflict resolution mechanisms within the WUAs and their capacities to fairly resolve disputes would be very important for social cohesion and increased cooperation for collective action.

An attempt was also made to examine rule conformance as an indicator of the level of cooperation for local management of irrigation schemes. Prompt payment of water levies and conformance to water allocation schedules were analyzed. Analysis of the survey data shows that the WUAs depend largely on water levies to carry out maintenance activities which involve monetary outlay. Fee collection performance was generally above average, with a mean payment rate of 70.8%. Indeed, the payment rate ranges from 0 in 17.3% of the WUAs to 100 in 42.3% of the user groups.

Results of a Probit estimation of cooperation (levy payment) and noncooperative behavior (water poaching) show that institutions play an important role in inducing compliance. Local rules such as payment of fines and forfeiture of rights to farm have a positive and significant impact on cooperative behavior, and significantly reduce the probability of water poaching. Consistent with rational behavior, default rates and water poaching are high in groups where deviant behavior receives only a warning. This finding underscores the importance of legal backing that would make institutions designed by the groups for the management of irrigation schemes more credible and durable. Improvements in irrigation services (water supply) resulting from quality maintenance offer incentives for rule conformance. Water shortages are disincentivizing to levy payment. The study also finds that the level of cooperation is higher where there are other social organizations. Other organizations are a proxy for social capital in the community and that social capital is associated with higher rule conformance.

Social cohesiveness and social relationships also matter in the success of community-based resource management. Articulation between users appears to be increasingly problematic where the number of villages using schemes increases. Indeed, the results of the study suggest that rule conformance is difficult to achieve when multiple villages share the use of the same schemes.

This could be due to high cost of monitoring behavior and the difficulty of enforcing norms across villages. Rule conformance is also high where bylaws are seen to have been designed by the WUA, but not imposed from outside. Consistent with other studies on collective action on local commons, group size is found to be very important for assuring cooperative behavior. Rule violation is higher in larger groups, thus supporting theoretical predictions that cooperative behavior is more likely to occur and be sustained in smaller groups. In smaller user groups peer monitoring is easier, shared norms and patterns of reciprocity are more common, and social sanctions are easier to implement through reputation mechanisms. Significantly, the application of labor intensive techniques for the rehabilitation of the schemes stimulate rule conformance and for that matter cooperation for successful local management of the irrigation schemes.

The results also highlighted the detrimental effect of the interference of landlords in land distribution on cooperation. Resistance of landlords to land redistribution weakens institutional mechanisms within the WUA, making enforcement of rules difficult. As these landlords allocate land to their favorites, they not only create factionalism within the user groups but also results in shifting alliances that become detrimental to collective action for the maintenance of the irrigation systems. Inequality in plot allocation lowers cooperation and encourages rule violation. The ability of the WUAs to tackle local asymmetries to promote equity and devise appropriate institutions that create incentives for all is important to improve cooperation for the sustainable management of the schemes.

The results also conclude that rule violation is prevalent in schemes where the rotational system of water allocation is used and in larger schemes. However, there is less probability of rule violation when the water allocation rule is crafted by the community. Thus, the choice of appropriate distributional arrangement and effective monitoring regimes can significantly reduce water poaching in larger schemes.

As a measure of the determinants of success of community management, the study has tried to examine the factors that explain the differences in maintenance performance across the schemes.

The results of the analyses show that quality of maintenance is lower in larger groups, suggesting that cooperation for successful local management is more likely to be achieved and sustained in smaller groups. Frequent training of group leaders on operation and maintenance techniques is very essential for improved quality of maintenance. Quality of maintenance is high in profitable schemes, as higher returns induce greater participation by group members in collective maintenance activities.

Socio-cultural (ethnic) homogeneity also promotes collective action and for that matter efficient management of the schemes. Indeed, quality of maintenance is higher where villages sharing the use of a scheme are of the same ethnic identity. Resistance of landlords to land redistribution has a negative effect on the quality of maintenance. Apart from undermining the authority of the WUA, the noncooperative attitudes of the landlords negatively affect group cohesion and unity required for successful collective action.

Quality of maintenance is low in communities where rules are frequently violated. However, local regulatory mechanisms elicit cooperation for collective action in the management of the irrigation schemes. Credible threat of sanctions such as fines and forfeiture of one right to farm for not contributing to maintenance are associated with high quality maintenance. In particular, monitoring and sanctioning limit free-riding and provide incentives for members to fulfill their maintenance obligations. This supports the views in the literature that suggest that without functioning institutions common-pool resources will not be effectively managed.

An attempt is also made to investigate the factors that affect the incentive for households to contribute efforts to the maintenance of the community irrigation schemes. Our results suggest that household labor availability plays a significant role in determining the level of household contribution to the maintenance effort. Households with more adult members are likely to have less labor constraints and for that matter can contribute more labor to the maintenance of the schemes. The results also highlight the important role profitability plays in household attitude to maintenance. Households that receive relatively higher returns contribute more effort to the maintenance of the schemes.

The multiple functions of the irrigation schemes in water scarce communities make the sustainability of the schemes a concern for the majority of the households. Household participation in maintenance activities is very high in schemes that experience frequent water shortages. Also of significant effect is household satisfaction with the quality of services they receive, particularly irrigation water schedules. Members who do not get water at the right time and in the required amount resort to digging of shallow wells. This reduces their dependence on the schemes and affects commitment to its maintenance.

Both external wages and market access have strong negative effects on effort contribution to maintenance. While higher wages increase the opportunity cost of maintenance time, market integration generates exit options that reduce the role of irrigation in households' livelihood strategies. Higher wages (and in general, exit options) outside the schemes increase the opportunity cost of labor and reduce the incentive for households to participate in the maintenance of the irrigation schemes. This presents a potential conflict between irrigation maintenance and poverty-alleviation strategies that promote off-farm economic activities.

Transparent and accountable leadership is shown to be an important concern that affects the incentives for households to contribute to the maintenance of the schemes. Lack of transparency and accountability and incidences of rent-seeking reduce trust and confidence in leadership, and undermine management efficiency. Leaders perceived to be corrupt lose their moral authority to enforce rules and regulations. The envisaged registration of the WUAs as cooperative organizations and the subsequent application of cooperative norms of periodic auditing of accounts of the association could help to improve accountability and reduce its negative effect on cooperation within the groups.

Two other outcomes that have important implications for the on-going programs to rehabilitate broken down schemes, in the study area, are perceptions about the quality of work done by the contractors engaged for the rehabilitation of the schemes, and community participation in the rehabilitation process. Households whose members participated in the program contribute more labor to the maintenance of the schemes. Physical participation in the rehabilitation program bestows on the participants some skills and a sense of ownership and moral commitment to contribute to the operation and maintenance of the schemes. Perceptions about the quality of the rehabilitation work also have a significant effect on household maintenance participation, as poor quality of works conducted weakens morale. In communities where dams were perceived to have been poorly rehabilitated, deteriorations are often attributed to the poor quality work done by the contractors; the irrigators tend to wait and hope that the contractor is brought back to rectify the anomaly, while the level of deterioration grows worse.

Consistently, the results of the study show that local regulatory mechanisms play an important role in ensuring the success of collective action for the maintenance of the community irrigation schemes. Graduated sanctions are significantly associated with high levels of maintenance participation and rule conformance. Rules and practices concerning participation as well as a credible threat of sanctions deter free riding. It is essential that farmers enforce bylaws and regulations, and strengthen monitoring and evaluation mechanisms to ensure efficient and sustainable management. Policy measures that facilitate the development of local institutions could enhance the ability of local organizations to sustainably manage their resources. Enactment of legislation recognizing the WUAs as the legitimate users and managers of the irrigation systems will go a long way to promote legitimacy and efficiency of the local organizations.

7.2   Conclusions

In this study we have attempted to analyze the factors that determine the success of collective action for local management of irrigation schemes in northern Ghana. To do this, we examined the level of cooperation and the factors that account for the differences in the quality of maintenance across the schemes. Finally, we examined the factors that motivate household participation in the collective maintenance of the irrigation schemes.

The findings suggest that accountable leadership, a high perceived quality of the rehabilitation work, physical participation of users in the rehabilitation process, higher benefits (larger plot sizes to assure higher returns), and satisfaction with water distribution schedules promote participation in collective action for the maintenance of the irrigation schemes.

Communities achieve greater cooperation for successful collective maintenance of the schemes where the user group size is small, fewer villages share the use of an irrigation scheme, and land redistribution policy was not resisted. Outside wages and market access can have detrimental effects on the success of collective action for the maintenance of the community irrigation schemes, as exit options distract commitments for the sustainability of the schemes.

The use of labor intensive methods for the rehabilitation of the schemes promotes group cohesion, a sense of ownership as well as moral commitment to ensure the sustainability of the resource. A greater sense of ownership by members of the user group stimulates improved monitoring and system maintenance. Credible threats of sanctions and rules crafted by the users themselves have higher marginal effects of promoting cooperation. Also, ethnic heterogeneity and inequality in plot allocation can be detrimental to collective action. The ability of the WUA to tackle local asymmetries to promote equity and devise appropriate incentives for all its members is essential for achieving a sustainable irrigation management.

7.3 Policy recommendations

Devolving rights to local communities to manage resources, and to establish and enforce use rules and regulations is, however, only a necessary condition for success of community management of natural resources. Sustainable resource management requires that rules and regulations of the user group be effectively observed. This will be difficult to achieve if the user groups lack legal backing. The durability of local institutions designed for the management of the schemes depends on their general acceptance and enforceability. Incorporating WUA norms into district assembly bylaws will not only legitimize the status of the WUA but also increase the ability of the user groups to enforce their rules and regulations.

There should be greater emphasis on training of irrigation groups on the key tasks of irrigation management and to demonstrate the benefits of community participation in irrigation management. The training programs should also provide the farmers with some skills of institution building and organizational management to ensure high competence and commitment in the execution of their duties.

Devolving operation and management activities of the schemes to users groups alone cannot guarantee quality maintenance and for that matter the sustainability of the schemes. There is the need for GIDA (with the participation of the user groups) to undertake periodic inspection of the schemes to ascertain the state of the infrastructure and to advise the WUAs on maintenance requirements.

Though difficult to address through policy measures, devolution programs must consider factors such as ethnic heterogeneity, especially of villages sharing use of a common resource. Federated structures or sub-group formation that place members of the same village in different groups with elected leaders could be helpful in reducing the negative effects of social heterogeneity on group performance.

As has been emphasized earlier, the labor intensive techniques for rehabilitation programs, though slow in implementation, have important implications for the sustainability of the schemes. Physical involvement in the rehabilitation process promotes sense of ownership and moral responsibility to participate in maintenance activities to ensure the sustainability of the schemes. Where project deadlines do not favor the use of the technique, efforts should be made to recruit more of the members of the beneficiary communities as laborers to ensure their active involvement in the process. Moreover, it was shown that perceptions about the quality of rehabilitation work done by the contractors affect the long-term sustainability of the schemes. It is important that monitoring and evaluation is strengthened to ensure that contractors engaged in the rehabilitation process deliver on designed specifications.

Trust is very vital for group cohesion. Encouraging the WUAs to register as cooperative organizations will help to streamline financial management in the associations while periodic auditing of accounts will promote accountability and improve trust in the user groups.

Resistance to land redistribution not only undermines the equity and poverty alleviation objectives of devolution programs, but also deprives the user groups of the level of cooperation needed for successful collective management. The project management should either accelerate efforts to legally acquire the command areas for the use of the irrigators associations or facilitate the acquisition of the land by the user groups. This will make it easier for women and the underprivileged groups to gain equal access to secured irrigation land, and to grant the users groups greater control over the management of the schemes.

The study tried to analyze possible determinants for successful collective action for the management of irrigation schemes in northern Ghana. It is hoped that the results of this study will contribute to the understanding of the determinants of success of collective action on local commons.

REFERENCES

Abatania, L.N. and Albert, H. 1993. Potential and constraints of legume production in the farming systems of Northern Ghana. In Proceedings of the *Third Workshop on Improving Farming Systems in the Interior Savanna Zone of Ghana*, Nyankpala Agricultural Experimental Station, NEAS, Tamale, Ghana. pp 170-181.

Abdulai, A. and Delgado, C. 1999. Determinants of nonfarm earnings of farm-based husbands and wives in northern Ghana. *America Journal of Agricultural Economics*, 81(1):117-130

Abdulai, A. and Regmi, P.P. 2000. Estimating Labor Supply of Farm Households under Nonseparability: Evidence from Nepal. *Agricultural Economics* 22(3): 309-320.

Adams, W.M. 1991. Large scale irrigation in Northern Nigeria: Performance and ideology. *Transactions of the Institute of British Geographers* 16:287-300.

Agrawal, A. 2001. Common property institutions and sustainable governance of resources". *World Development* 29(10): 1649-1672.

Agrawal, A, and Yadama, G. 1997. How do local institutions mediate market and population pressures on resources? Forest Panchayats in Kumaon, India. *Development and Change* 28(3): 437-66.

Agrawal, A., and Goyal, S. 2001. Group size and collective action: Third party monitoring in common-pool resources. *Comparative Political Studies* 34(1): 63-93.

Agodzo, S.K., Osei, G., and Kranjac-Berisavljevic, G. 1998. *Some operation and maintenance issues in formal-sector irrigation in Ghana*. Paper presented at an International Conference in Agricultural Engineering, organized by the Ghana Institute of Engineers, held at University of Cape Coast, Ghana, September 20-25, 1998.

Agyepong, G.S. Gyasi, E.A., Nabila, J.S., and Kufogbe, S.K., 1999. Population, land –use and the environment in a West African savannah ecosystem: An Approach to sustainable land-use on community lands in northern Ghana. *In People and their Planet: Searching for Balance*, Baudo, B.S., and Moomaw, W.R. (eds.): pp 251-271. Macmillan Press Ltd.

Alesina, A., and La Ferrara, E. 2000. Participation in heterogeneous communities. *Quarterly Journal of Economics* 115(3): 847-904.

Amonor, K.S. 1994. *The new frontier: Farmer's response to land degradation. A West African study*. Ondon: Zed Books Limited.

Andanye, J.E., and Sadiq, A. 1994. Turning over of small scale irrigation schemes to farmers: a role of Ghana irrigation development authority. A paper presented at the First Regional Workshop on Farmer-management of Small Dams in the Upper East Region, held at IFCAT, Novrongo, 23-24 Sept. 1994.

Anyane, S. La. 1962, Agriculture in the general economy. In *Agriculture and Land Use in Ghana*. Bryn Wills (ed): 192-200. Oxford University Press,.

Ardeifio-Shandorf, E. 1982. Rural development strategies in northern Ghana: Problems and prospects for reaching the small farmer. Occasional Paper 18/1982. Center for Development Studies, Swansea, Norwich.

Arnold, J.E.M. 1998. *Managing forests as common property*. Forestry Paper 136/1998. Food and Agricultural Organization (FAO), Rome.

Arnold, J.E.M., and Campbell, J.G. 1986. Collective management of hill forests in Nepal: The community forestry development project. In *Proceedings of the Conference on Common Property Resource Management*. National Research Council: 425-454. Washington, D.C: National Academy Press.

Appiah-Opoku, S. and Hyma, B. 1999. Indigenous institutions and resource management in Ghana. *Indigenous Knowledge and Development Monitor*, 7(3) pp15-17.

Appiah-Opoku, S. and Mulamoottil, G. 1997. Indigenous institutions and environmental assessment: the case f Ghana. *Environmental Management* 21(2): 159-171.

Apter, D. 1968. *Ghana in transition*. New York: Macmillan Company

Asenso-Okyere, W. K., Twum-Baah, K. A., Kasanga, A., Anum, J., and Pörtner, C. (2000). Ghana living standards survey. Report of the Fourth Round (GLSS 4). Ghana Statistical Service. Accra:

Asfaw, A. 2003. Cost of illness, demand for medical care, and the prospects of community health insurance schemes in the rural areas of Ethiopia. *Development Economics and Policy*. Heidhues, F. and von Braun, J. (Series eds):Vol. 34. Frankfurt, Peter Lang.

Asiamah, R.D. 2002. Plinthite development in upland agricultural soils in Ghana. In *Proceedings of World Congress of Soil Science Symposium* 22, Paper no. 1448:1-12.

Axelrod, R. 1984. *The evolution of cooperation*. Basic Books, New York.

Ayariga, G.A. 1992. *Small scale irrigation in northern Ghana: A Modus operandi for farmer management*. Unpublished MSc Thesis, Cranfield Institute of Technology, Silsoe College.

Ayariga, G.A.R. 1993. Guidelines for promoting farmer-management of small dams in the Upper East Region of Ghana. Working paper I – *Guidelines for promoting the formation of water users associations*, Upper East Land Conservation and Smallholder Rehabilitation Project (LACOSREP), Bolgatanga.

Ayariga, G.A.R. 1995. *The LACOSREP water users' association programme. Revised strategies and current status*. Working Paper II. MOFA/IFAD, Bolgatanga, Ghana.

Ayariga, G.A.R. 2002. *Determinants of success of community-based irrigation development in the Upper East Region of Ghana: Case studies from the LACOSREP program*. Mimeo, Center for Development Research (ZEF), Bonn.

Bacho, F.Z.L. 2001. *Infrastructure delivery under poverty: Portable water provision through collective action in northern Ghana.* Spring Research Series 34.

Bakang, J.A. and C.J. Garforth. 1998. Property rights and renewable natural resources degradation in North-Western Ghana. *Journal of International Development* 10: 501-514.

Baker, J.M. 1997. Common property resource theory and Kuhl irrigation systems of Himachal Pradesh, India. *Human organization* 56: 199-208.

Baland, J.M., and Platteau, J.P. 1996. *Halting degradation of natural resources: is there a role for the rural communities?* Oxford: Claredon Press.

Baland, J.M., and Platteau, J.P. 1997. Wealth inequality and efficiency in the commons Part I: the unregulated case. *Oxford Economic Papers* 49(3): 451-82.

Baland, J.M., and Platteau, J.P. 1998. "Wealth inequality and efficiency in the commons Part II: the regulated case." *Oxford Economic Papers* 50(1): 1-22.

Baland, J., and Platteau, J. 1999. The ambiguous impact of inequality and on local resource management. *World Development* 27(5): 773-788.

Baland, J.M, and Platteau, J.P. 2002. *Economics of common property management regimes.* Centre de Recherche en Economie du Dévelopment (CRED) Department of Economics, Faculty of Economics, Business, and Social Sciences, University of Namur, Belgium. February 6, 2002 version.

Bardhan, P.K. 1993. Analytics of institutions of informal cooperation in rural development. *World Development* 21(4): 633-639.

Bardhan, P. 2000. Irrigation and cooperation: An empirical analysis of 48 irrigation communities in South India". *Economic Development and Cultural Change* 48(4): 847-865.

Bardhan, P. 2001. Distributive conflicts, collective action, and institutional economics. In G.M. Meier and J.E. Stiglitz (eds), *Frontiers of development economics: The future perspective.* Oxford University Press.

Benneh, G. 1973. *Land tenure and farming systems in a Sissala village in northern Ghana.* Bulleting de I ' IFAN 2, Dakar, pp 361-379.

Benning, R.B. 1996, Land Ownership, divestiture and beneficiary rights in Northern Ghana:- Critical issues. In Seminar Report on *Decentralisation, Land Tenure and Land Adminstration in Nothern Ghana.* Accra: Konrad-Adenauer-Stiftung.

Blaylock, J.R. and W.N. Blisard. 1994. Women and the demand for alcohol: Estimating participation and consumption. *Journal of Consumer Affairs* 27: 319-334.

Berkes, F. (ed) 1989. *Common property resources: Ecology and community-based sustainable development.* London: Belhaven Press.

Berkes, F., and Folke, C. (eds). 1998. *Linking social and ecological systems: Management practices and social mechanisms for building resilience.* Cambridge: Cambridge University Press.

Binswanger, H.P., and McIntyre, J. 1987. Behavioral and material determinants of production relations in land-abundant tropical agriculture. *Economic Development and Cultural Change*, Vol. 36(1): 73-99.

Blaylock, J.R., Blisard, W.N. 1992. U.S. Cigarette Consumption: The case of low-income women. *American Journal of Agricultural Economics* 74(3): 698-705.

Blundell, R., and Meghir, C. 1987. Bivariate alternative to the Tobit model, *Journal of Econometrics*, 34:179-200.

Braimoh, A.K. 2004. Modeling land-use change in the Volta Basin of Ghana. *Ecology and Development Series*. Vlek, P.L.G. et al (Series eds), No. 14. Göttingen: Cuvillier Verleg.

Brewer, J. et al. (1999), *Irrigation Management Transfer in India: Policies, Processes and Performance*. Indian Institute of Managements, Ahmedabad, India, and the IWMI, Colombo, Sri Lanka. Oxford and IBH Publishing Co. Pvt. Ltd.

Bromley, D., (ed.). 1992. *Making the Commons Work: Theory, Practice and Policy*. San Francisco: Institute for Contemporary Studies Press.

Bromley, D.W., and Cernea, M.M. 1989. *The management of common property natural resources: Some conceptual and organizational fallacies*. World Bank Discussion Papers, 5/1989. The World Bank, Washington, D.C.

Bruce, J., Kibuuka, M.A., Neupane, K.P., Okomoda, J.K., Phonekhampheng, O., and Puskur R. 1999. *Towards sustainable agricultural development: Research and development options for improved integration of crop-livestock-fishery systems in irrigated and rainfed areas of the Upper East Region of Ghana*. ICRA Working Document Series 77, Ghana-1999.

Buchanan, J.M. 1965. An economic theory of clubs. *Economica*, 32: 1-14.

Caldwell, J. and Caldwell, P. 1987. The cultural context of high soil fertility in Sub-Saharan Africa. *Population and Development Review*, vol. 13(3): 409-438.

Campbell, B., Mandandom, A., Nemarundwe, N. Jong, W., Luckret, M. and Matose, F. 2001. Challenges to proponents of common property resource systems: Despairing voices from the social forests of Zimbabwe. *World Development* 24 (4): 589-600.

Cernea, M. 1988. User groups as producers in participatory afforestation strategies. World Bank staff Discussion Papers No. 70. Washington, D.C., The World Bank.

Chamberlin, J. 1974. Provision of collective goods as a function of group size. *American Political Science Review* 68(2): 707-16.

Chambers, R. 1988. *Managing canal irrigation: Practical analysis from South Asia*. Delhi: Oxford University Press.

Ciriacy-Wantrup, S.V., and Bishop, R.C. 1975. Common property as a concept in natural resource policy. *Natural Resources Journal* 15(4):713-728.

Cleaver, F. 1998. Moral ecological rationality, institutions and the management of communal resources. Presented at "Cross Boundaries", the 7th annual

Conference of the International Association for Study of Common Property, British Columbia, Canada, June 10-14, 1998.

Cones, R., and Sandler, T. 1986. *The theory of externalities, public goods, and club goods*. Second edition

Coser, L.A. 1956. *The functions of Social Conflicts*. Glencoe: Free Press

Clottey, V.A., and Kombiok, J. 2000. Land use types in the Northern Region of Ghana. A paper presented at the Workshop on Land Use Types in the Volta Basin, GILLBT, Tamale, Nov. 13-15.

Cragg, J.G. 1971. Some statistical models for limited dependent variables with application to the demand for durable goods. *Econometrica*, 39:829-844.

Cutler, S.J. 1976. Age differences in voluntary association membership. *Social Forces* 35(1): 43-58.

Dasgupta, P. 1993. *An Inquiry into well-being and destitution*. New York: Oxford University Press.

Davidson, R. and MacKinnon, J.C. 1993. *Estimation and inference in econometrics*. New York: Oxford University Press.

Dayton-Johnson, J. 1999. Irrigation organizations in Mexican *unidades de riego*: Results of a field study. *Irrigation and Drainage Systems* 13(1): 57-76.

Dayton-Johnson, J. 2000a. Determinants of collective action on the local commons: a model with evidence from Mexico. *Journal of Development economics* 62(1): 181-208.

Dayton-Johnson, J. 2000b. Choosing rules to govern the commons: A model with evidence from Mexico. *Journal of Economic Behavior and Organization* 42(1): 19-41.

Dayton-Johnson, J. and Bardhan, P.P. 2002. Inequality and conservation on the local commons: A theoretical exercise. *Economic Journal* 112(481): 459-703.

Demestz, H. 1967. "Towards a theory of property rights". *American Economic Review* 57(2): 347-359.

Dietz, T., Dolsak, N., Ostrom, E., and Stern, P.C. 2002. The drama of the commons. In E. Ostrom et al. (eds) *The Drama of the Commons*, Committee on Human Dimensions of Global Change, National Research Council, Washington, D.C., National Academy Press.

Dickson, K.B. and Benneh, G. 1988. *A new geography of Ghana*, Harlow: Longman.

Dinar, A., Rosegrant, M.W., and Meinzen-Dick, R. 1997. Water allocation mechanisms – principles and examples. World Bank Policy Research Working Paper 1779/1997. The World Bank, Washington, D.C.

Dittoh, S. 2000. *Agricultural land use agreements and derives rights for granting access to farm land in Northern Ghana*. Draft Report, DIFID, London.

Dogbe, E.G.K. 1998. Report on initiation workshop on savanna resource management project for Opinion Leaders in the Upper East Region, 9th-11th Sept. 1998.

Dove, M.R. 1993. A revisionist view of tropical deforestation and development. *Environment Conservation* 20: 17-24.

Davidson, R. and MacKinnon, J.G. 1993. *Estimation and inference in econometrics.* New York: Oxford University Press.

Dutilly-Diane, C., Sadoulet, E., and de Janvry, A. 2003. Household behavior under market failures: How natural resource management in agriculture promotes livestock production in the Sahel. *Journal of African Economies* 12(3): 343-370.

Easter, K.W., and Palanisami, K. 1986. Tank irrigation in India and Thailand: An example of common property resource management". Department of Agricultural Economics staff working paper, University of Minnesota.

Edig, A. van, Engel, S. and Laube, W. 2002. Ghana's water institutions in the process of reforms: From the international to the local level. In Neubert, S., Scheumann, W., and van Edig (eds.) *Reforming Institutions for Sustainable Water Management.* German Development Institute (DIE), Bonn. Germany.

Elhorst, J.P. 1994. Firm-households interrelationships on Dutch dairy farms. *European Review of Agricultural Economics* 21: 259-276.

Engel, S. 2002. "To share or not to share: A model of land conflicts and irrigation management in Ghana." Center for Development Research, University of Bonn, Germany.

Engel, S. 2003. *Endogeneities in participatory resource management: Political economy aspects of irrigation management in Ghana.* Jahrestagungs des Ausschusses für Entwicklungsländer des Vereins fur Socialpolitik, Humburg, 4-5, Juli, 2003.

Eyre-Smith, R. St J. 1933, *A brief history and social organization of the people of the Northern Territories of the Gold Cost.* Accra.

FAO (Food and Agriculture Organization). 1982. Farmer's participation and organization for Irrigation Water Management. Land and Water Development Division, FAO, Rome.

FAO. 1989. Consultation on irrigation in Africa. *Irrigation and Drainage Paper* 42/1989. FAO/UNDP.

Fearson, J.D., and Laitin, D.D. 1996. Explaining interethnic cooperation. *American Political Science Journal* 90(4): 715-735.

Florkowski, W.J., Moon, W., Resurreccion, A.V.A, Paraskova, P. Beuchat, L.R. Murgov, K., and Chinnan, M.S. 2000. Allocation of time for meal preparation in transition economy. *Agricultural Economics* 22(2): 173-183.

Fujita, M. Hayami, Y., and Kikuchi, M. 1999. The conditions for collective action for local commons management: The case of irrigation in the Philippines. Paper presented at World Bank Seminar. Washington, D.C.

Garcia-Barrios, R., and Garcia-Barrios, L. 1990. Environmental and technological degradation in peasant agriculture: A consequence of development in Mexico. *World Development* 18(11):1569-1585.

Gibson, C. 2001. Forest resources: Institutions for local governance in Guatemala. In E. Burger, E. Ostrom, R.B. Norgaard, D. Policansky, and B.D.

Goldstein (eds), *Protecting the commons: A framework for natural resource management in the Americas*. Washington, D.C.: Island Press.

Gibson, C. C., and Becker, C.D. 2000. A lack of institutional demand: Why a strong local communities in Western Ecuador fails its forest. In C.C. Gibson, M.A. Mckean, and E. Ostrom (eds), *People and forests: Communities, institutions, and governance*. Cambridge, MA: MIT Press.

Ghana, 2002. Agenda for growth and prosperity. Analysis and policy statements. *Ghana poverty reduction strategy, 2002/2004*, Accra.

Greene, W.H. 2003. *Econometric analysis*. Fifth Edition. New Jersey: Prentice Hall, International Edition.

Groothuis, P.A. and Miller, G. 1994. Locating hazardous waste facilities: The influence of NIMBY beliefs. *American Journal of Economics and Sociology* 53 (3): 335-347

Groenfeldt, D. 1997. Transferring Irrigation Systems from the State to users: Questions of Management, Authority, and Ownership. Paper presented at the 96th Annual meetings of the American Anthropological Association, Washington, DC, November 19, 1997.

GSS (Ghana Statistical Service). 2000, *Poverty Trends in the 1990s*. Accra.

GSS, 2000. Ghana living standards survey (GLSS) IV. Accra.

GSS. 2002, *2000 Population and Housing Census: summary of Report of Final Results*. Accra

Gyasi, K.O. 2001. Integration of Maize Markets in Northern Ghana. In Badu-Apraku B., M.A.B. Fakorade, M. Ouedrago, and R.J. Casky (eds) *Impact, Challenges and Prospects of Maize Research in West and Central Africa*. Proceedings of Regional Maize Network 4-7 May, 1999. IITA-Cotonou, Benin. Pages 433-446

Gyasi, K.O., L. Abatania, A.S. Langyintuo, and P. Terbobri, 2002. Determinants of adoption of improved rice varieties in the inland valleys of northern Ghana: A Tobit model application. *SADAOC Review of Studies on Food Security in Central West Africa* 1(1): 4-26.

Gyasi, K.O., and Engel, S. 2002. Community-based irrigation management in northern Ghana. Summary of Proceedings of Workshop on Participatory Irrigation management in Northern Ghana, held in Bolgatanga, Ghana, May 8, 2002.

Gyasi, K.O., Engel, S., and Frohberg, K. 2004. Determinants of household incentive to participate in collective action: The case of community-based irrigation management in northern Ghana. A Paper Presented at the Fifth Workshop on Institutional Analysis, Barcelona, 25 June 2004

Hardin, G. 1968. The dilemma of the commons, *Science* 162(3859): 1243-1248.

Hardin, R. 1982. *Collective action*, Baltimore, MD: John-Hopkins University Press.

Hactett, S.C. 1992. Heterogeneity and the provision of governance for common-pool resources. *Journal of Theoretical Politics* 4:325-342.

Hayami, Y., and Ruttan, V. W. 1985. *Agricultural development: An international perspective.* Baltimore: John-Hopkins University Press.

Hayami, Y., and Kikuchi, M. 1982. *Asian village economy at the cross road.* John-Hopkins University Press.

Heckman, J.J. 1979. Sample selection Bias as a specification error. *Econometrica* 47:153-161.

Heltberg, R. 2001. Policy Options: Determinants and impacts of local institutions for common resource management. *Environment and Development Economics* 6(2): 183-208.

Hesse, J.H. 1998. *The sustainability of animal traction farming systems in northern Ghana.* Eschborn: GTZ.

Hirshleifer, D., and E. Rasmusen. 1989. Cooperation in a repeated prisoner's dilemma with ostracism. *Journal of Economic Behavior and Organization* 12(1): 87-106.

Hotes, F.L. 1982. The World irrigation experience. *Development Policy Review* 5: 99-123.

IFAD (International Fund for Agricultural Development). 1999. Report and Recommendations of the President to the Executive Board on the Proposed Loan to the Republic of Ghana for the Upper East Region Land conservation and Smallholder Rehabilitation Project (LACOSREP)– Phase II. Rome, April 1999.

IFAD. 2003. *A brief institutional assessment of water users' association in northern Ghana: Early stages of pro-poor local institutional development in irrigated smallholder agriculture.* A draft paper prepared by Ian Jones and Norman Messer.

Isaac, R.M., Walker, J., and Williams. 1993. Group size and voluntary provision of public goods: Experimental evidence utilizing large groups. *Journal of Public Economics* 54(1): 1-36

Jacoby, H.G. 1992. Productivity of Men and women and the Sexual Division of Labor in Peasant Agriculture of Peruvian Sierra. *Journal of Development Economics* 37(2): 265-287

Jacoby, H.G. 1993, Shadow wages and peasant family labor supply: an econometric application to the Peruvian Sierra. *Review of Economic Studies* 60(4): 903-921.

Jayaraman, T.K. 1981. Farmers organizations and surface irrigation projects: Two empirical studies from Gujarat. *Economic and Political Weekly* 16 (39).

Jensen, M, and Ostrom, E. 2001. Critical factors that foster local self-governance of common-pool resources: the role of endogeneity. Paper presented at a conference on *Inequality, Collective Action, and Environmental Sustainability*, Santa Fe Institute, New Mexico, USA September 21-23.

Joesch, J.M, and Hiedemann, B.G. 2002. The demand for nonrelative child care among families with infants and toddlers: a double-hurdle approach. *Journal of Population Economics* 15(3): 495-526.

Johnson, R.N. and Libecap. 1982. Contracting problems and regulations: the Case of the fisheries. *American Economic Review* 72(2): 1005-23.

Johnson, O.E.G. 1972. Economic analysis, the legal framework and land tenure systems. *Journal of Law and Economics*, Vol. 15: 259-276.

Johnson, S.R. and G.C. Rausser. 1971. Effects of misspecifications of log-linear functions when values are zero or negative. *American Journal of Agricultural Economics*, 53, 120-124.

Jones A.M. 1989. A double-Hurdle Model of Cigarette Consumption. *Journal of Applied Econometrics* 4(1):23-39.

Jones, A.M. and J.M. Labeaga. 2003. Individual heterogeneity and censoring in panel data estimates of tobacco expenditure. *Journal of Applied Econometrics*, vol. 18(2): 157-177.

Kamara, A. 2001. Property rights, risk and livestock development in Ethiopia. *Socio-economic Studies on Rural Development, Vol. 123.* Wissentschaftsverlag Vauk, Kiel.

Kamara, A., van Koppen, B., and Magingxa, L. 2001. Economic viability of small-scale irrigations systems in the context of state withdrawal: The Arabie scheme in the Northern Province of South Africa. 2nd WARSFA Waternet Symposium: Integrated Water Resource Management: Theory, Practice, Cases. Cape Town, 30-31 October.

Kanbur, R. 1992. *Heterogeneity, distribution and cooperation in commons property management.* Background paper for the 1992 world development report. World Bank.

Kasanga, R.K. 1994. *Land tenure systems and ecological degradation in Northern Ghana. What role for local authorities?* London: The Royal Institute of Chartered Surveyors.

Kasanga, R. K. 1996. The role of chiefs and 'tendamba' in land administration in northern Ghana. In Seminar Report on *Decentralisation, Land Tenure and Land Adminstration in Nothern Ghana.* Accra: Konrad-Adenauer-Stiftung

Kikuchi, M., and Hayami, Y. 1980. Inducements to institutional innovations in an agrarian economy. *Economic Development and Cultural Change*, Vol. 29(1): 21-36.

Kim, O. and Walker, M., 1984. The free rider problem: Experimental evidence. *Public Choice*, 43(1), 3-24.

Kimber, R. 1981. Collective action and the fallacy of liberal fallacy. *World Politics*, Vol. 32(2): 172-196.

Kingma, B. 1989. An accurate measure of crowd-out effect, income effect, and price effect for charitable contribution. *Journal of Political Economy*, 97:1197-1207.

Kipo, T. 1993. Farmer's perception of the factors of production and agriculture in northern region. In *Proceedings of the Third Workshop on Improving Farming Systems in the Interior Savanna Zone of Ghana*, Nyankpala Agricultural Experimental Station, NEAS, Tamale, Ghana, pp 189-197.

Kiss, A. 1990. *Living with the wildlife: wildlife resource management with local participation in Africa.* World Bank Technical Paper 130/1990, Washington, D.C: The World Bank.

Knox, A., and Meinzen-Dick, R. 2001. *Workshop on collective action, property rights and devolution of natural resource management: Exchange of knowledge and implication for policy.* A Workshop Summary Paper. CAPRI Working Paper 11/2001, Washington, D.C: IFPRI.

Koning, P. 1996, *The state and rural class formation in Ghana: A comparative analysis.* London: KPI Ltd.

Korem, A. 1985. *Bush Fire and Agricultural Development in Ghana.* Accra: Ghana Publishing Cooperation.

Kranjac-Berisavlyevic, G. 2000, Irrigation as a measure against desertification in Northern Ghana. Unpublished Paper. University for Development Studies, Tamale.

Kranz, B., W-D. Fugger, J. Kroschel, and J. Saurborn. 1998. The influence of organic manure on *Striga hermonthica* (Del.) Benth infestation in northern Ghana. In, H.P. Blume, H. Egger, E. Fleischhauer, a. hebel, C. Reij, K.C. steiner (eds.) Towards Sustainable Land Use: Further cooperation between people and institutions. *Advances in GeoEcology* 31(1): 615-619.

Kreps, D.M., Milgrom, P., Roberts, J., and R. Wilson. 1982. Rational cooperation in the finitely repeated prisoner's dilemma. *Journal of Economic Theory* 27(2): 245-52.

Lam, W.F. 1998. *Governing irrigation systems in Nepal: Institutions, infrastructure, and collective action.* San Francisco: Institute for Contemporary Studies.

Larson, B.A., and D.W. Bromley. 1990. Property rights, externalities and resource degradation: Locating the tragedy. *Journal of Development Economics* 33(2): 235-262.

Lawry, S.W. 1990. Tenure policy toward common property natural resources in Sub-Saharan Africa". *Natural Resources Journal* 30 (Spring): 403-422.

Lee, L.F. 1982. Some approaches to the correction of selectivity bias. *Review of Economic Studies* 49: 355-372

Levi, M. 1988. *Of rule and revenue.* Berkeley: University of California Press.

López-Nicolás, A. 1998. Unobserved heterogeneity and censoring in the demand for health care. *Health Economics* 7(5): 429-437.

López, R.E. 1984. Estimating labor supply and production decisions of self-employed farm producers. *European Economic Review* 24(1): 61-82.

MaCurdy, T.E and J.H. Pencavel. 1986. Testing between competing models of wage and employment determination in unionized markets. *Journal of Political Economy* 94:S3-S39.

Maddala, G.S. 1983. *Limited-dependent and qualitative variables in econometrics.* Cambridge: Cambridge University Press

Maier-Rigaud, Frank, 2000. "Under what conditions are decentralized solutions to collective action problems likely?". In *Constituting the Commons*. Eighth Annual Conference of IASCP, Bloomington, Indiana, USA, May 31 – June 4.

Makombe, G, Meinzen-Dick, R., Davies, S.P., and Sampath, R.K. 2001. An evaluation of Bani (Dambo) systems of smallholder irrigation development strategy in Zimbabwe. *Canadian Journal of Agricultural Economics* 49: 203-216

Mas-Colell, A., Whinston, M., and Green, J. 1995. *Microeconomic Theory*, Oxford: Oxford University Press.

Magorian, C. 1992. *Participation and hazardous waster facility sitting: An annotated bibliography*. Philadelphia Environmental Research Council.

Manukiam, M. 1952. *The Tribes of the Northern Territories of the Gold Coast*, London: International African Institute.

Marwell, G., and Oliver, P.E. 1993. *The critical mass in collective action: A micro-social theory*. Cambridge: Cambridge University Press.

McCarthy, N., A.B. Kamara, and M. Kirk. 2003. Cooperation in risky environment: Evidence from Southern Ethiopia. *Journal of African Economies* 12 (2): 236-270.

McCarthy, N., and J-P. Vanderlinden. 2002. *Climatic variability and cooperation in rangeland management: A case study from Niger*. CAPRi Working Paper 24/202. Washington, D.C: International Food Policy Research Institute (IFPRI).

McCarthy, N., Sadoulet, E. and de Janvry, A. 2001. Common pool resource appropriation under costly cooperation. *Journal of Environmental Economics and Management* 42(3): 297-309.

McCay, B.J. and Acheson, J.M. 1990. The question of the commons: The culture of ecology of communal resources. Tuscon: University of Arizona Press.

Meinzen-Dick, R. 1997. Farmer participation in irrigation: 20 years of experience and lessons for the future. *Irrigation and Drainage Systems*.

Meinzen-Dick, R.S. and Bakker, M. 2001.Water rights and multiple uses. *Irrigation and Drainage Systems* 15(2): 129-148.

Meinzen-Dick, R., Raju, K.V., and Gulati, A. 2002. What affects organization and collective action for managing resources? Evidence from canal irrigation systems in India. *World Development*, Vol. 30(4): 649-666.

Meinzen-Dick, R., and Knox, A. 1999. *Collective action, property rights, and devolution of natural resource management: A conceptual paper*. Paper Presented at the International Workshop on Collective Action, Property Rights and Devolution of Natural Resource Management: Exchange of Knowledge and Implication for Policy. Puerto Azul, Philippines, June 21-25.

Mensah-Bonsu, A. 2003. *Migration and Environmental Pressure in Northern Ghana*. Ph.D. Thesis, Vrije Universiteit Amsterdam.

Merry, D.J. 1996. *Institutional Design principles for accountability in large scale irrigation systems*. IWMI Research Report.

MoFA (Ministry of Food and Agriculture). 1998. *Facts and figures.* Policy Planning, Monitoring, and Evaluation Department (PPMED), MoFA, Accra.

Murdoch, J.C., Sandler, T., and Sargent, K. 1997. A tale of two collectives: sulphur versus nitrogen emissions reduction in Europe. *Economica*, 64(254): 281-302.

Murdoch, J.C., T. Sandler, and W.P: Vijverberg. 2003. The participation decision and level of participation in an environmental treaty: a spatial probit analysis. *Journal of Public Economics*, 87(2): 337-362.

Myers, N. 1991. The World's forest and human populations: The environmental interconnections. In K. Davies, and M. Bernstam (eds.), *Resources, Environment, and Population: Present Knowledge, Future Options.* New York: Oxford Publishing Press.

Nelson, R.R. 1995. Recent evolutionary theorizing about economic change. *Journal of Economic Literature*, vol. XXXIII: 480-490.

North, D.C. 1981. *Structure and Change in Economic History.* New York: Norton.

Nugent, J.B. 1993. Between state, markets and households: A noninstitutional analysis of local organizations and institution. *World Development* 21(4): 623-632.

Nyari, B. S. 2002. Report on diagnostic survey of the landholding system by the water users associations at the rehabilitated dam sites under LACOSREP I. MoFA/IFAD Upper East Land Conservation and Smallholder Rehabilitation Project, Bolgatanga, Ghana.

Oakerson, R.J. 1986. A model for the analysis of common property problems. In *Proceedings of the conference on common property resource management.* National Research Council. Washington, D.C: National Academy Press.

Oakerson, R.J. 1992. Analyzing the commons: A framework. In Bromley (ed) *Making the commons work: Theory practice and policy*, San Francisco: ICS Press.

Okamura, M. 1991. Estimating the impact of the Soviet threat on the United States-Japan alliance. *Review of Economics and Statistics*, 73(2): 200-207.

Ostrom, E. 1997. Self-governance and common-pool resources. W97-2. Workshop in Political Theory and Policy Analysis, Indiana University, Bloomington.

Ostrom, E. 1999. *Self-governance and forest resources.* Occasional Paper 20/1999. Bogor, Indonesia, CIFOR.

Ostrom, E. 1992a. *Crafting institutions for self-governing irrigation systems.* San Francisco: Institute of Contemporary Studies.

Ostrom, E. 1992b. The rudiment of a theory of origins, survival, and performance of common property institutions. In, *Making the Commons work: Theory, Practice and Policy.* Bromley (ed). San Francisco: Institute for Contemporary Studies Press.

Ostrom, E. 1990. *Governing the commons: The evolution of institutions for collective action.* Cambridge: Cambridge University Press.

Ostrom, E. 1997. Self-governance of common-pool resources. W97-2. Workshop in Political Theory and Analysis. Indiana University, Bloomington.

Ostrom, E., R. Gardner, J. Walker. 1994. *Rules, games and common-pool resources*. Ann Arbor, Mich.: University of Michigan Press.

Ostrom, E., and R. Gardner. 1993. Coping with asymmetries in the commons: Self-governing irrigation systems can work. *Journal of Economic Perspectives* 7(4): 93-112.

Pender J. and Scherr S. 1999. *Organizational development and natural resource management: Evidence from central Honduras*. Environment and Production Technology Division Discussion Paper 49/1999. Washington, DC: IFPRI (International Food Policy Research Institute).

Poteete, A., Ostrom, E., 2003. Institutional mediation of group characteristics and the consequences for collective action. Working Paper No. W03-12. Workshop in Political Theory and Policy Analysis. Indiana University, Bloomington, IN.

Poulton, C. 1998. Cotton production and marketing in Northern Ghana: The dynamics of competition in a system of interlocking transactions. In *Smallholder cash crop production under market liberalization: A new institutional economics perspective*. Dorward, A., J. Kydd, and C. Poulton (eds):56-112. UK: CAB International.

Puhani, P.A. 2000. The Heckman correction for sample selection and its critique. *Journal of Economic Survey* 14(1): 53-68.

Putnam, R. 1993. *Making democracy wok: Civic traditions in modern Italy*. Princeton University Press.

Rangan, H. 1997. Property versus control: The state and forest management in the Indian Himalaya. *Development and Change* 28(1): 71-94.

Rappaport, R.A. 1984. *Pigs for the ancestors: Rituals in the ecology of a New Guinea people*. New Haven, CT: Yale University Press

Rasmussen, L-N., and Meinzen-Dick, R. 1995. *Local organizations for natural resource management: lessons from theoretical and empirical literature*. ETPD Discussion Paper 11/1995. Washington, D.C., IFPRI.

Reidinger, R.B. 1980. Water management by administrative procedures in an Indian irrigation system. In W.E. Coward (ed9, *Irrigation and Agricultural Development in Asia*, Itheca, USA: Cornell University Press.

Reuben, R. 2003. *The evolution of theories of collective action*. www.tinbergen.nl/~reuben/colaction.pdf Accessed in August 2003.

Ribot, J.C. 1999. Decentralization, participation, and accountability in Sahelian forestry: Legal instruments of political-administrative control. *Africa* 69(1): 23-65.

Richards, M. 1997. Common property resource institutions and forest management in Latin America". *Development and Change* 28(1): 95-117.

Ringer, C., Rosegrant, M, and Paiser, M. 2000. *Irrigation and water resources in Latin America and the Caribbean: Challenges and strategies*. EPTD Discussion Paper No. 64. Washington, D.C: IFPRI.

Runge, C.F. 1986. Common property and collective action in economic development. *World Development*, Vol. 14 No. 5: 623-635.

Runge-Metzer, A. 1993. Farm household systems in northern Ghana. In *Farm households systems in northern Ghana: A case study in farming systems oriented research for the development of improved crop production systems*. A. Runge-Metzer, and L. Diehl (eds). Weikersheim, Germany.

Ruthenberg, H. 1980. *Farming systems in the tropics*, Third Edition. Oxford.

Salas, M.A. 1994. The Technicians only believe in science and cannot read the sky. In

Sam-Amoah, L.K. and T.W. Gowing. 2001. The experience of irrigation management transfer in Ghana: A case study of Dawhenya irrigation scheme". *Irrigation and Drainage Systems*. 15(1): 21-38.

Sandler, T. 1992. *Collective Action, Theory and Application*. Ann Arbor: Michigan University Press

Seabright, P. 1993. Managing local commons: Theoretical issues in incentive design. *Journal of Economic Perspectives* 17(4): 113-134.

Sengupta, N. 1992. *Managing common property: Irrigation in India and the Philippines*. New Delhi: Sage.

Shaffer, J.D. 1969. On institutional obsolescence and innovation – Background for professional dialogue and policy. *American Journal of Agricultural Economics*, pp. 245-267.

Shah, T., von Koppen, B., Merrey, D., de Lange, M., and Samad, M. 2002. *Institutional alternatives in African smallholder irrigation: Lessons from international experience in irrigation management transfer*. IWMI Research Report 60. International Water Management Institute, Colombo, Sri Lanka.

Singh, I., Squire, L., and Strauss, J. 1986 (eds). Agricultural Household Models. John-Hopkins University Press.

Skoufias, E. 1993. Seasonal labor utilization in agriculture: The theory and evidence from agrarian households in India. *American Journal of Agricultural Economics*, 75(February): 20-32.

Skoufias, E. 1994. Using Shadow Wages to Estimate Labor Supply of Agricultural Households. *American Journal of Agricultural Economics* 76 (May): 215-227.

Smith, R. 1981. Resolving the tragedy of the commons by creating private property rights in wildlife. *CATO Journal* 1: 439-468.

Smith, D.H. 1994. Determinants of voluntary association participation and volunteering: A literature review. *Nonprofit and Voluntary Sector Quarterly* 23(3):243-263.

Soones and Thompson (eds) *Beyond farmer first. rural people's knowledge, agricultural research and extension practice*. London: IT Publications.

Spradly, J.P., and McCurdy, D.W. 1980. *Anthropology: The cultural perspective*, New York: John Wiley & Sons.

Stern, N.A., and Edwards, V.M. 1998. Platforms for collective action in multiple-use common pool resources. Presented at "Cross Boundaries", the 7th annual Conference of the International Association for Study of Common Property, British Columbia, Canada, June 10-14, 1998.

Stevenson, G.G. 1991. Common property economics: A general theory and land use application. Cambridge University Press.

Sugden, R. 1986. *The economics of rights, co-operation and welfare.* Blackwell, Oxford.

Svendsen, M., Trava, J. and Johnson, S.H. III. 1997. Participatory irrigation management: benefits and second generation problems. Lessons from an international works held at CIAT, Cali, Colombia, 9-15 February.

Swendsen, M. and Nott, G. (1997), Irrigation management transfer in Turkey: Early experience with a national program under rapid implementation. IWMI, Colombo, Sri Lanka.

Tang, S.Y. 1992. *Institutions and collective action: Self-governance in irrigation.* San Francisco, institute of Contemporary Studies.

Thijssan, G.J. 1992. *Microeconomic models of Dutch dairy firms.* Wageningen Economic Studies, No. 23. Wageningen Agricultural University.

Tripp, R.B. 1982. Time allocation in Northern Ghana: An example of random visit method. *Journal of Development Areas*, 10: 391-400.

Tural, H. 1995. Recent trends in irrigation management changing direction for the public sector. *Natural Resource Perspective*, No. 5.

Underhill, H. 1990. *Small scale irrigation in Africa in the context of rural development.* Cranfield Press.

Uphoff, N. 1986a. *Local institutional development: an analytical sourcebook with cases.* West Hartford, CT: Kumarian Press.

Uphoff, N. 1986b. *Improving international irrigation management with farmer participation: Getting the process right.* Boulder, Colo., USA: Westview Press.

Uphoff, N. 1992. *Local institutions and participation for sustainable development.* Gatekeepers series, No.3. London: International Institute for Environment and Development.

Upholf, N. and Langholz, J. 1998. Incentives for avoiding the tragedy of the commons. *Environmental Conservation* 25(3): 251-261.

Ward, R.W. and Moon, W. 1995. Evaluating beef checkoff programs: alternative approach. In, *Economic Analysis of Meet Promotion*. Kinnucan, H.W., Lenz, J.E., Clary, C.R. (eds): 67-78. Ithaca: Cornell University Press

Warren, D.M. 1991. Using Indigenous Knowledge in Agricultural Development. World bank Discussion Paper 12/1991. Washington D.C: World Bank.

Weissings, F., and Ostrom, E. 1990. Irrigation institutions and the games irrigators play: Role of enforcement without guards. In *Game equilibrium models*, vol.2. R. Salter (ed). Berlin: Springer.

White, H. 1980. A heteroskedasticity-consistent covariance matrix estimator and a direct test for heteroskedasticity. *Econometrica*, Vol. 48(4): 817-828.

White, T.A., and Runge, C.F. 1994. Common property and collective action: lesion from comparative watershed management in Haiti. *Economic Development and Cultural Change* 43(1):1-41.

White, T.A., and Runge, C.F. 1995. The emergence and evolution of collective action: Lessons from of watershed management in Haiti. *World Development* 23(10): 1683-1693.

Willis, J.B. (ed), 1962. *Agriculture and land use in Ghana*. Oxford University Press.

Woldehanna, T. 2000. *Economic analysis and policy implications of farm and off-farm employment: a case study in the Tigray region of Northern Ethiopia*. PhD Dissertation, Wageningen University.

World Bank. 1987. The Upper Region Agricultural Development Program (URADEP). End of Project Report. Washington, D.C: World Bank.

World Bank. 1986. Sector-Sector Review Report. Ghana Irrigation Development Authority. Washington D.C: World Bank.

Vaidyanatha, A. 1984. Water control institutions and: A comparative perspective. *Indian Economic Review*, 20: 25-84.

Van der Linden, J.P. 1999. Conflicts and cooperation over the commons: A conceptual and methodological framework for assessing the role of local institutions. In *Proper rights, risks and livestock development in Africa*. McCarthy, N., B. swallow, M. Kirk and P.Hazel (eds): 351-370. Washington, D.C: IFPRI.

Varughese, G. and Ostrom, E. 2001. The contested role of heterogeneity in collective action: Some evidence from community forestry in Nepal. *World Development* 29(5): 747-765.

Vermillion, (1994). Irrigation Management Transfer: Towards an Integrated Management Revolution. Address given at the International Conference on Irrigation Management Transfer, Wuhan, China, September 20-24, 1994

Vermillion, D.L. 1997. *Management Devolution and the Sustainability of irrigation: Results of Comprehensive versus Partial Strategies*. Paper presented at the FAO/World Bank Technical Consultation on Decentralization and Rural Development, 16-18 December, Rome.

Vermillion, D. 1999. *Proper rights and collective action in the devolution of irrigation system management*. Paper presented at Workshop on "Collective Action, Property Rights, and Devolution of natural Resources", Puerto Azul, Philippines, June 21-24.

Wade, R. 1987. The management of common property resources: Collective action as an alternative to privatization or state regulation. *Cambridge Journal of Economics* 11(1): 95-106.

Wade, R. 1988. *Village republics: Conditions for collective action in South India*. Cambridge: Cambridge University Press.

Weissing, F., and E. Ostrom. 1991. Irrigation institutions and the games irrigators play: Rule enforcement without guards. In *Game equilibrium models II: Methods, morals, and markets*. R. Selten (ed.). Springer Verlag, Berlin.

Yen, S.T. 1995. Cross-section estimation of US demands for alcoholic beverage. *Applied Economics*, 26:381-392.

Yen, S.T. and H. Jensen. 1995. *Determinants of expenditure on Alcohol*. Working Paper 144/1995. Center for Agricultural and Rural Development. Iowa State University.

Yen, S.T. 2003. Estimating demand for cigarette and alcohol with zero observations: a censored system. Selected paper presented at the AAEA annual meeting, Montreal, Quebec, Canada, July 2003.

Yoder, R. 1981. Non-agricultural uses of irrigation systems: Past experiences and implications for planning and design. Paper prepared for the Agricultural Development Council, Inc.

Yoder, R. 1994. Locally managed irrigation systems: Essential tasks and implication for assistance, management transfer and turnover program. International Irrigation Management Institute, Colombo, Sri Lanka.

Young, K.R. 1994. Roads and the environmental degradation of tropical montane forests. *Conservation Biology* 8(4): 972-976.

Young, O.R. 1995. The problem of scale in human and environment relationships. In *Local commons and global interdependence: Heterogeneity and cooperation in two domains*. R.O. Keohane amd E.Ostrom (eds): 27-46. London: Sage Publications.

Zak, P.J., and Knack, S. 2001. Trust and growth. *The economic Journal* 111(4): 295-321.

APPENDICES

Appendix A
Table A1: Location of community irrigation schemes

Code	Region	District	Community dam	Latitude	Longitude
1	UER	Bolgatanga	Winkongo	10° 42.644N	0° 51.637W
2	UER	Bolgatanga	Bolga Soe	10° 47.833N	0° 50.772W
3	UER	Bolgatanga	Puso-Namongo	10° 39.694N	0° 51.000W
4	UER	Bolgatanga	Baare	10° 44.224N	0° 47.736W
5	UER	Bongo	Bongo central	10° 54.503N	0° 47.421W
6	UER	Bongo	Gambrongo	10° 53.512N	0° 54.154W
7	UER	Kasena-Nankena	Doba	10° 51.611N	1° 02.038W
8	UER	Kasena-Namkena	Tellania	10° 55.662N	1° 03.537W
9	UER	Bolgatanga	Sumbrungu	10° 49.732N	0° 56.208W
10	UER	Kasena-Nankena	Goo	10° 52.950N	1° 06.196W
11	UER	Kasena-Nankena	Saboro	10° 54.877N	1° 05.332W
12	UER	Kasena.Nankena	Gia	10° 54.692N	1° 08.172W
13	UER	Kesena-Nankena	Paga-Nania	10° 58.989N	1° 06.720W
14	UER	Builsa	Bandem	10° 39.878N	1° 16.922W
15	UER	Builsa	Chiok	10° 40.535N	1° 14.869W
16	UER	Builsa	Sinyansa	10° 37.894N	1° 16.410W
17	UER	Builsa	Wiaga	10° 39.887N	1° 16.067W
18	UER	Bawku-West	Kamenga	10° 50.912N	0° 30.912W
19	UER	Bawku-West	Binaba II	10° 46.606N	0° 28.555W
20	UER	Bawku-West	Kusanaba	10° 44.125N	0° 28.768W
21	UER	Bawku-West	Binaba I	10° 45.422N	0° 28.255W
22	UER	Bawku-West	Saka	10° 57.730N	0° 26.040W
23	UER	Bawku-West	Teshie (Zebilla II)	10° 57.664N	0° 28.178W
24	UER	Bawku-West	Tanga	10° 54.393N	0° 26.376W
25	UER	Bawku-West	Tonde	10° 53.200N	0° 26.169W
26	UER	Bawku-West	Googo	11° 01.320N	0° 24.871W
27	UER	Bawku-West	Lamboya	10° 54.982N	0° 31.410W
28	UER	Bawku-West	Tilli	10° 53.206N	0° 32.106W
29	UER	Bawku-East	Kuka Central	11° 00.773N	0° 10.895W
30	UER	Bawku-East	Bugri	10° 58.439N	0° 08.109W
31	UER	Bawku-East	Gagbiri	10° 54.677N	0° 07.762W
32	UER	Bawku-East	Worikambo	10° 45.474N	0° 07.864W
33	UER	Bawku-East	Woriyanga	10° 53.317N	0° 04.110W
34	UER	Bawku-East	Basiyonde	10° 59.259N	0° 00.960W
35	UER	Bawku-East	Bingouri	10° 59.689N	0° 18.301W
36	UER	Bawku-East	Kpalwega	11° 02.044N	0° 16.398W
37	UER	Bawku-East	Anissi	10° 57.361N	0° 16.888W
38	UER	Bawku-East	Binduri	10° 57.860N	0° 18.576W
39	UER	Bawku-East	Sakpari	10° 56.860N	0° 20.553W
40	UER	Bawku-East	Kumplagogo	10° 57.325N	0° 20.052W
41	UER	Bawku-East	Sabzunde	10° 59.886N	0° 02.999W
42	UER	Bawku-East	Nafkulga	10° 53.387N	0° 21.866W
43	UER	Bawku-East	Kaadi	10° 55.735N	0° 18.320W
44	UWR	Wa	Busa	10° 00.666N	2° 23.177W
45	UWR	Nadowli	Sankana	10° 11.084N	2° 36.249W
46	UWR	Jirapa/Lambusie	Kaane	10° 39.861N	2° 37.764W
47	UWR	Jirapa/Lambusie	Han	10° 14.089N	2° 14.089W
48	UWR	Lawra	Babile	10° 32.905N	2° 51.217W
49	UWR	Sissala	Bulu	10° 52.406N	2° 16.074W

50	UWR	Sissala	Welembelle	10° 30.672N	1° 57.582W
51	UER	Bongo	Adaboya	10° 54.244N	0° 43.844W
52	UWR	Bongo	Dua	10° 53.253N	0° 46.827W

Appendix B

Formula for calculating volume of water available for irrigation

To estimate the volume of water available for irrigation we adopt the formula used by the LACOSREP to analyze the storage capacities of its dams. The following assumptions are made based on the analysis of LACOSREP dams:
1. The 'dead storage' volumes average 2% of the total storage capacities (full supply level).
2. Loss through seepage and environmental losses estimated at 38% of full supply level.
3. Animal and domestic uses 5% of full supply level.[44]

Let total volume at full supply level		=	V
Losses:			
Seepage and evaporative losses	= L	=	38%V
Animal and domestic requirements	= AU	=	5%V
Dead storage	= D	=	2%V
Total water available for irrigation		=	V – (L +AU + D)
		=	55%V

Based on the above formula the crop water requirement, for instance, for 0.8 hectare of vegetable and 0.2 hectare of maize grown at the dam site in the dry season gives volumes of water required as 8,535m3 for vegetables and 2,308m3 for maize making a total of 10,844m3 which has been rounded of to 11,000m3/ha. So the amount of land that can be irrigated with the available water is 55%V1 / 11000.

---

[44] Source: J.E. Andanye, LACOSREP, Bolgatanga

Appendix C
Profitability indices (Gross margins and Cost-Benefits ratios)

The potential of the irrigation system to generate enough income to satisfy the income expectations of the irrigators is a greater incentive for households to participate in the collective maintenance of the scheme. Efficient management, it is agued, influences the productivity and profitability both at plot and scheme levels. At the scheme level viability depends on the ability to organize collective action for the maintenance of the schemes, make and enforce regulations. We estimated the financial benefits and costs of the community based schemes to determine their profitability, using data from 2002/2003 dry season gardening at the schemes. Using actual cost of production, market prices at the time of harvest, and plot yield data (extrapolated to per ha bases), returns and cost budgets were developed for each scheme. Gross margins achieved were computed as follows:

$GM = Gross\ Return - Cash\ Cost$

$Gross\ margin\ per\ unit\ of\ water = GM/1000m^3$

Benefit Cost ratios were calculated as follows
$BC = GM/cash\ cost$

Cash cost include direct cash cost associated with vegetable production in he dry season; value of seed, fertilizer, pesticides, manure, and hired labor. It does not include depreciation of equipment and direct and indirect cost of scheme maintenance.

Gross returns consist of the value, in Cedis, of all crops households produced at the irrigation schemes (and sold) during the 2002/2003 dry season. These include the value of crops kept for seed, given to persons as present or in-kind payments, home consumption and value of the quantity sold.

Table C1 summarizes the estimated profitability indices for the 52 small-scale schemes under communal management. The profitability indices were estimated from 2002/2003 dry season gardening production data collected as part of the farm household survey we conducted. It must be emphasized however that 2002/2003 dry season was greatly affected by a combination of disease complex that results in crop losses in many of the irrigation schemes in the Upper East Region. The crops that were seriously affected were tomatoes and pepper.

Table C1: Profitability indices (Mean gross margin and Benefit-Cost ratios)

Community scheme	Mean Gross margin/ha	B - C ratio	C.V	Community scheme	Mean Gross margin/ha	B - C ratio	C.V[a]
Winkongo	18758884.93	3.86	1.55	Lamboya	756637.88	16.08	44.91
Bolga soe	7328739.80	2.29	1.07	Tilli	31595546.88	5.99	0.92
Puso-Namongo	1031949.25	1.44	1.51	Kuka central	130678026.30	4.81	2.00
Baare	4854344.18	2.52	1.66	Bugri	21491038.24	29.43	2.63
Bongo Central	9066608.50	3.93	1.36	Gagbri	38385522.06	6.73	1.03
Gambrongo	3518961.83	4.76	2.32	Worikanbo	41688895.38	6.75	1.70
Doba	518837.50	1.38	4.19	Woriyanga	44078493.05	5.08	1.24
Tellania	14577150.55	2.72	1.46	Basiyonde	65413882.35	4.60	1.13
Sumbrungu	11711049.17	1.64	2.18	Binguri	48180926.84	6.73	0.68
Goo	17559792.71	7.66	0.82	Kpalwega	29070433.29	6.23	0.57
Saboro	9111547.38	3.46	1.75	Anissi	9520109.75	4.17	0.81
Gia	9697530.00	5.98	0.99	Binduri	44007651.13	9.65	1.58
Paga-Nania	24845815.99	2.02	1.79	Sakpari	46449637.82	9.28	0.81
Bandem	24881991.13	7.75	1.98	Kumpalgogo	102819934.20	12.00	2.09
Chiok	12608646.08	2.77	1.86	Sabzunde	36640256.05	5.63	1.62
Sinyansa	26279162.50	22.41	0.82	Nafkulga	9094118.01	2.57	0.88
Wiaga	47425146.05	3.03	2.39	Kadi	8332912.81	4.22	0.90
Kamenga	50690261.21	5.35	1.34	Busa	18887726.20	8.83	1.78
Binaba II	23206695.67	6.93	0.93	Sankana	5748030.25	39.17	2.26
Kusanaba	98803691.31	10.10	2.91	Karni	8350330.13	7.12	0.84
Binaba I	26566290.25	5.54	1.32	Heng (Han)	30044515.95	7.66	1.06
Saka	21810924.83	4.66	1.31	Babile	43738387.18	9.33	1.31
Teshie	31136068.75	5.07	1.48	Bollu	21856201.33	6.98	0.97
Tanga	20835891.38	3.50	1.03	Wellembele	9526636.73	8.71	0.97
Tonde	60441305.50	7.09	2.60	Adaboya	1039269.00	3.71	2.51
Googo	339150750.00	8.30	1.15	Dua	10634304.94	9.61	0.85

[a] Coefficient of variation of gross margin among farmers within the schemes.

Appendix D

Table D4.1: Age group versus level of education attained

		Level of Education Attained[*]					
		No formal education	Primary	JSS[1]/ Middle	SSS[2]/ Secondary	Post secondary	Total
Age Group	=< 25	25	15	15	12	2	69
	25-35	78	16	14	18	2	128
	36-45	76	24	21	9	1	131
	46-55	60	5	17	0	4	86
	56-65	52	2	6	1	1	62
	>= 65	43	0	1	0	0	44
Total		334	62	74	40	10	520

[1]. JSS = Junior Secondary School: Three-year system of basic education (after primary school) that has replace the 4-year middle school education after Ghana's education reform in 1987
[2]. SSS= Senior Secondary School: Three-year secondary education that has replaced the 7-year secondary education in the old system
[*]The correlation coefficient of age versus years of education = -0.265 (0.000, 2-tailed).

Table D6.1: OLS estimates of profitability

	Coef	S.E.	t	P>| t		95% conf. interval	
Input price index[+]	0.06325	1.03986	0.06	0.952	-2.0450	2.1715
Output price index[+]	0.69714	0.79429	0.88	0.386	-0.9137	2.3080
Forfeiture	0.73740	0.36075	2.04	0.048	0.0057	1.4690
Rule Conformance*	0.20977	0.24775	0.86	0.395	-0.2846	0.7041
WUA size	0.00168	0.00064	2.62	0.013	0.0003	0.0029
Water shortage	-0.30542	0.25758	-1.19	0.244	-0.8278	0.2169
Employment	0.21710	0.26097	0.83	0.411	-0.3121	0.7463
Ethnicity	0.77047	0.34732	2.22	0.033	0.0660	1.4748
Villages	-0.01367	0.04773	-0.29	0.776	-0.1104	0.0831
Market access	0.08923	0.27966	0.32	0.752	-0.4779	0.65641
Hetero index	1.50905	0.70832	2.13	0.040	0.0733	2.9465
Social org.	0.30905	0.13538	2.28	0.028	0.0344	0.5836
Maintenance*	0.78839	0.44906	1.76	0.088	-0.1223	1.6991
Plot gini	-0.14183	0.94259	-0.15	0.881	-2.0525	1.7698
Constant	10.5549	1.49749	7.05	0.000	7.5178	13.592

Dependent variable Ln (Gross margin)
Sample size =52; F (15, 36) = 2.52; Prob. > F = 0.0118; R^2 = 0.5118; Adjusted R^2 = 0.3084
Root MSE = 0.7896

+ See Appendix E for formula output quantity index ad output price index
*[1] Proxy: Community-wide rule against non-participation in communal activities
*[2] Proxy: Community satisfied with quality of rehabilitation

Table D6.2: OLS estimates of quality of maintenance

	Coef	Robust SE	t	P>| t		95% conf. interval	
WUA size	-0.00038	0.00035	-1.09	0.282	-0.0011	0.0003
Plot gini	-1.71957	1.80448	-0.95	0.347	-5.3867	1.9475
Plot gini squared	2.48408	2.25079	1.10	0.277	-2.0900	7.0582
Training	0.41635	0.19316	2.16	0.038	0.0237	0.8089
Forfeiture of plots	0.43615	0.17947	2.43	0.021	0.0714	0.8008
Fine	0.29256	0.15426	1.90	0.066	-0.0209	0.6060
Profitability*	0.42157	0.17224	2.45	0.020	0.0715	0.7716
Water shortages	0.05807	0.13580	0.43	0.678	-0.2179	0.3340
Labor intensive technique	0.74366	0.17969	4.14	0.000	0.3784	1.1088
Ethnicity of villages	0.29795	0.15819	1.88	0.068	-0.0235	0.6194
Market access	0.00449	0.13889	0.03	0.974	-0.2778	0.2867
Hetero index	0.87461	0.32184	2.72	0.010	0.2205	1.5286
Rule conformance*[1]	-0.15001	0.17675	-0.85	0.402	-0.50921	0.2092
Wage rate	-0.00019	0.00010	-1.88	0.069	-0.0004	0.00001
Quality of rehabilitation	-0.23089	0.13286	-1.74	0.091	-0.5008	0.0391
Social interaction	0.14495	0.07541	1.92	0.063	-0.0083	0.2982
Landlord resist land redistribution	-0.23212	0.21405	-1.08	0.288	-0.6662	0.2038
Constant	-2.94647	0.55050	-5.35	0.000	-4.0652	-1.8277

Dependent variable: LN(Quality index)
Sample size = 52; F(17 , 34) = 3.92; Prob. > F=0.0003; R^2 = 0.6620; Adjusted R^2 = 0.4930; Root MSE =0.3583
*[1]Predicted value of water poaching.

Table D6.3: Tobit estimates of levy payment performance

	Coef	Robust SE	t	P>| t		95% conf. interval	
Fine	29.73634	15.02155	1.98	0.055	-0.7002	60.1728
Forfeiture of plots	41.61716	21.97554	1.89	0.066	-2.9095	86.1438
Maintenance*	0.01906	0.00800	1.36	0.181	-0.0053	0.02712
WUA size	-0.08363	0.048288	-1.73	0.092	-0.1814	0.0142
Water shortages	9.28859	13.01109	0.71	0.480	-17.0743	35.6515
Conflicts*	9.38964	25.52684	0.37	0.715	-42.3326	61.1119
Market access	-12.99805	12.73234	-1.02	0.314	-38.7962	12.8001
Social org.	28.45496	19.3118	1.47	0.149	-10.6744	67.5843
Hetero index	-2.25395	36.70166	-0.06	0.951	-76.6185	72.1106
Plot gini	-81.92207	45.01633	-1.82	0.077	-173.133	9.2896
Socinter	0.63821	0.53544	1.19	0.241	-0.4466	1.72312
Landlords resist	-16.96530	26.67282	-0.64	0.529	-71.0095	37.0789
Labor intensive	26.21737	14.18759	1.85	0.073	-2.5294	54.9641
Profitability*	16.06886	13.78514	1.21	0.233	-11.2302	44.6324
Distribution rule*	48.06886	20.56437	2.34	0.025	6.4014	89.7363
Constant	-214.23660	185.90900	-1.15	0.257	-590.924	162.4507
se	36.48496	4.142229				

Dependent variable: Levy payment performance (percent of who have paid levy)
Sample size = 52 ; LR χ^2 (15) = 22.25; Prob. > χ^2 = 0.1014; Pseudo R^2 = 0.0471

Table D6.4: Cobb-Douglas production function estimation of vegetable production

	Coef	Robust SE	t	P>| t |	95% conf. interval	
LnFertilizer	0.09573	0.02315	4.13	0.000	0.0502	0.1412
LnChemicals	0.02007	0.01048	1.92	0.056	-0.0005	0.0406
LnTomato seeds	0.01505	0.01605	0.94	0.349	-0.0164	0.0465
LnOnion ``	0.06407	0.02253	2.84	0.005	0.1979	0.1083
LnPepper ``	0.06285	0.01817	3.46	0.001	0.0271	0.0985
LnOkra ``	0.02835	0.01776	1.60	0.111	-0.0065	0.0632
LnLettuce ``	0.03137	0.02693	1.16	0.245	-0.0215	0.0842
LnBean ``	0.03581	0.04584	0.78	0.435	-0.0542	0.1258
Ln(Trad. Leafy Veget.) ``	-0.02374	0.01453	-1.63	0.103	-0.0523	0.0048
Ln(Fix inputs)	0.08525	0.03688	2.31	0.021	0.0127	0.1577
Ln(Maintenance labor)	0.01498	0.01999	0.79	0.454	-02430	0.0542
Ln(Household labor)	0.17065	0.04436	3.85	0.000	0.0834	0.2578
Ln(Hired labor)	0.01915	0.01381	1.39	0.166	-0.0079	0.0463
Ln(Share high value crops)	1.24636	0.20945	5.95	0.000	0.8348	1.6578
Ln(Share of irrig. land)	0.06719	0.37520	1.87	0.061	-0.0032	0.1376
District dummies						
Bolgatanga	0.06395	0.37520	0.17	0.865	-0.6732	0.8011
Bongo	0.25919	0.39635	0.65	0.513	-0.5195	1.0379
Kasena-Nankena	-0.00238	0.38174	-0.01	0.995	-0.7524	0.7476
Builsa	-0.11004	0.39951	-0.28	0.783	-0.8942	0.6742
Bawku West	0.42183	0.39933	1.06	0.291	-0.3862	1.2064
Bawku East	0.75344	0.40292	1.87	0.062	-0.0382	1.5450
Wa	0.42577	0.43203	0.99	0.325	-0.4230	1.2746
Jirap/Lambusie	0.40338	0.44557	0.91	0.336	-0.4720	1.2788
Lawra	0.38407	0.49262	0.78	0.436	-0.5838	1.3519
Sissala	0.67755	0.43097	1.57	0.117	-0.1692	1.5243
Constant	10.67210	0.60879	17.53	0.000	9.4959	11.868

Dependent variable Ln (Crop value)
Sample size = 520; F(25 , 494) = 33.28; Prob. > F = 0.0000; R^2 = 0.6228; Root MSE = 0.8550

Table D6.3: Tobit estimates of levy payment performance

	Coef	Robust SE	t	P>| t |
Age	-9.60524	7.53483	-1.27	0.203
Credit	466.4672	264.2439	1.77	0.078
Sex	849.4402	256.1892	3.32	0.001
Household size	41.6694	36.8882	1.13	0.259
School years	29.0406	28.4476	1.02	0.308
Badopp	-1117.951	237.0281	4.72	0.000
Distance to market	-19.6309	59.16586	-0.33	0.740
Off-farm income	119.9162	368.9537	0.33	0.745
Arable Land	38.3491	19.9199	1.93	0.055
Use other irrigation methods	724.8505	514.2503	1.41	0.159
Constant	-138.7846	540.054	-0.26	0.797

Sample size = 445; F (10, 434) = 4.25; Prob. > F = 0.000; R^2 = 0.067; Root MSE = 3029.1

Table D6.5: Determinants of level of participation in collective maintenance (with Rho)

Variable	Coefficient	Robust S.E	P >| t	
Constant	1.39111	0.46786	0.003
Age	0.02685	0.01072	0.012
Age Squared	-0.00026	0.00011	0.017
Household members < 15 years	0.00633	0.01296	0.625
Household members ≥15 years	0.03564	0.01018	0.001
Profitability	3.92817	1.84184	0.034
Off-farm income of household head	0.16804	0.07514	0.026
Forfeiture of plots	0.16889	0.10480	0.108
Land outside scheme	-0.00310	0.00588	0.598
Cont. labor for reh (labor intensive)	0.12143	0.07207	0.093
Satisfied with quality of service	0.10879	0.06275	0.084
Plot Gini	0.20478	0.22731	0.368
Accountable leaders	0.19098	0.13132	0.147
Distance to market	-0.13668	0.07920	0.085
Water shortage	0.13626	0.09610	0.157
Size of scheme	0.01519	0.00449	0.001
Age of the WUA	0.01741	0.01851	0.348
Shadow Wage per hour	-0.00016	0.00002	0.000
Social interaction	0.00623	0.00159	0.000
Plot is located at the head end	-0.00979	0.07701	0.599
Plot is located at the tail end	-0.04011	0.07594	0.598
Community is urban	0.01798	0.22562	0.939
Community is peri-urban	-0.03561	0.14286	0.803
Community is rural	-0.09242	0.10551	0.382
Inverse Mills ratio (RHO)	-0.04667	0.28621	0.871

No. of obs. = 445; $F_{(24, 420)}$ = 7.98; Pro > F = 0.0000; R-squared = 0.4459;
Root MSE= 0.653

Appendix E

Formula for calculating output quantity and output price index

Output price index (Po_i) for each scheme is calculated by dividing the value of total output by the volume of total crops.

The value of output is the sum of the physical yield of each crop (Y_j) multiplied by the a vector of associated local market prices (P_L), and divided by the plot area (A_i) under crop production Thus,

$$Value = \sum_{j=1}^{J} P_L Y_j \Big/ A_i$$

The volume of output is the sum of the physical yield of each crop (Y_j) multiplied by a vector of a constant market price (P_C), divided by the plot area (A_i) under crop production. Thus,

$$Volume = \sum_{j=1}^{J} P_C Y_j \Big/ A_i$$

Therefore output price index Po_i

$$Po_i = \frac{\sum_{j=1}^{J} P_L Y_j \Big/ A_i}{\sum_{j=1}^{J} P_C Y_j \Big/ A_i}$$

Development Economics and Policy

Series edited by Franz Heidhues and Joachim von Braun

Band 1 Andrea Fadani: Agricultural Price Policy and Export and Food Production in Cameroon. A Farming Systems Analysis of Pricing Policies. The Case of Coffee-Based Farming Systems. 1999.

Band 2 Heike Michelsen: Auswirkungen der Währungsunion auf den Strukturanpassungsprozeß der Länder der afrikanischen Franc-Zone. 1995.

Band 3 Stephan Bea: Direktinvestitionen in Entwicklungsländern. Auswirkungen von Stabilisierungsmaßnahmen und Strukturreformen in Mexiko. 1995.

Band 4 Franz Heidhues / François Kamajou: Agricultural Policy Analysis – Proceedings of an International Seminar, held at the University of Dschang, Cameroon on May 26 and 27 1994, funded by the European Union under the Science and Technology Program (STD). 1996.

Band 5 Elke M. Förster: Protection or Liberalization? A Policy Analysis of the Korean Beef Sector. 1996.

Band 6 Gertrud Schrieder: The Role of Rural Finance for Food Security of the Poor in Cameroon. 1996.

Band 7 Nestor R. Ahoyo Adjovi: Economie des Systèmes de Production intégrant la Culture de Riz au Sud du Bénin: Potentialités, Contraintes et Perspectives. 1996.

Band 8 Jenny Müller: Income Distribution in the Agricultural Sector of Thailand. Empirical Analysis and Policy Options. 1996.

Band 9 Michael Brüntrup: Agricultural Price Policy and its Impact on Production, Income, Employment and the Adoption of Innovations. A Farming Systems Based Analysis of Cotton Policy in Northern Benin. 1997.

Band 10 Justin Bomda: Déterminants de l'Epargne et du Crédit, et leurs Implications pour le Développement du Système Financier Rural au Cameroun. 1998.

Band 11 John M. Msuya: Nutrition Improvement Projects in Tanzania: Implementation, Determinants of Performance, and Policy Implications. 1998.

Band 12 Andreas Neef: Auswirkungen von Bodenrechtswandel auf Ressourcennutzung und wirtschaftliches Verhalten von Kleinbauern in Niger und Benin. 1999.

Band 13 Susanna Wolf (ed.): The Future of EU-ACP Relations. 1999.

Band 14 Franz Heidhues / Gertrud Schrieder (eds.): Romania – Rural Finance in Transition Economies. 2000.

Band 15 Katinka Weinberger: Women's Participation. An Economic Analysis in Rural Chad and Pakistan. 2000.

Band 16 Christof Batzlen: Migration and Economic Development. Remittances and Investments in South Asia: A Case Study of Pakistan. 2000.

Band 17 Matin Qaim: Potential Impacts of Crop Biotechnology in Developing Countries. 2000.

Band 18 Jean Senahoun: Programmes d'ajustement structurel, sécurité alimentaire et durabilité agricole. Une approche d'analyse intégrée, appliquée au Bénin. 2001.

Band 19 Torsten Feldbrügge: Economics of Emergency Relief Management in Developing Countries. With Case Studies on Food Relief in Angola and Mozambique. 2001.

Band 20 Claudia Ringler: Optimal Allocation and Use of Water Resources in the Mekong River Basin: Multi-Country and Intersectoral Analyses. 2001.

Band 42 Roukayatou Zimmermann: Biotechnology and Value-added Traits in Food Crops: Relevance for Developing Countries and Economic Analyses. 2004.

Band 43 F. Markus Kaiser: Incentives in Community-based Health Insurance Schemes. 2004.

Band 44 Thomas Herzfeld: *Corruption begets Corruption*. Zur Dynamik und Persistenz der Korruption. 2004.

Band 45 Edilegnaw Wale Zegeye: The Economics of On-Farm Conservation of Crop Diversity in Ethiopia: Incentives, Attribute Preferences and Opportunity Costs of Maintaining Local Varieties of Crops. 2004.

Band 46 Adama Konseiga: Regional Integration Beyond the Traditional Trade Benefits: Labor Mobility contribution. The Case of Burkina Faso and Côte d'Ivoire. 2005.

Band 47 Beyene Tadesse Ferenji: The Impact of Policy Reform and Institutional Transformation on Agricultural Performance. An Economic Study of Ethiopian Agriculture. 2005.

Band 48 Sabine Daude: Agricultural Trade Liberalization in the WTO and Its Poverty Implications. A Study of Rural Households in Northern Vietnam. 2005.

Band 49 Kadir Osman Gyasi: Determinants of Success of Collective Action on Local Commons. An Empirical Analysis of Community-Based Irrigation Management in Northern Ghana. 2005.

www.peterlang.de

Edilegnaw Wale Zegeye

The Economics of On-Farm Conservation of Crop Diversity in Ethiopia

Incentives, Attribute Preferences and Opportunity Costs of Maintaining Local Varieties of Crops

Frankfurt am Main, Berlin, Bern, Bruxelles, New York, Oxford, Wien, 2004.
XVIII, 243 pp., 9 fig.
Development Economics and Policy. Edited by Franz Heidhues and
Joachim von Braun. Vol. 45
ISBN 3-631-53142-7 / US-ISBN 0-8204-7378-2 · pb. € 45.50*

The issue of maintaining a diverse gene pool in the form of crop varieties is very topical world wide. This is caused by the potential benefit of crop genetic resources for addressing future demand emanating from unforeseen agricultural problems. This volume is mainly concerned with on-farm conservation as a supplement to the other in situ and ex situ conservation options. The study aims at generating relevant information for maintaining local varieties on farmers' fields in Ethiopia. In order to effectively devise policies for on-farm conservation, the volume argues that an improved understanding of farmers' incentives, attribute preferences and opportunity costs is indispensable. These issues are extensively addressed (both theoretically and empirically) with a focus on policy that is expressed by the guiding question *Given the socioeconomic set-up, what policy options are available to undertake on-farm conservation of crop diversity in Ethiopia?* The study results are intended to help identify optimal policies for on-farm conservation taking sorghum, coffee, and wheat as empirical examples.

Contentes: Overview of Variety Use and Crop Diversity Conservation in Ethiopia · Data Collection Methodology and Insights on Farmers' Variety Choice · Conservation of Crop Diversity and Economic Development: Issues, Paradigms and Economic Theories · Economic Analysis of Farmers' Incentives to Diversify on Local Sorghum Varieties · Farmers' Coffee Variety Attribute Preferences, Demand for Local Varieties and Incentives for Poly-Variety

Frankfurt am Main · Berlin · Bern · Bruxelles · New York · Oxford · Wien
Distribution: Verlag Peter Lang AG
Moosstr. 1, CH-2542 Pieterlen
Telefax 00 41 (0) 32 / 376 17 27

*The €-price includes German tax rate
Prices are subject to change without notice
Homepage http://www.peterlang.de